Lecture Notes in Physics

Edited by H. Araki, Kyoto, J. Ehlers, München, K. Hepp, Zürich
R. Kippenhahn, München, H. A. Weidenmüller, Heidelberg
and J. Zittartz, Köln
Managing Editor: W. Beiglböck

251

W0051161

R. Liebmann

Statistical Mechanics of Periodic Frustrated Ising Systems

Springer-Verlag
Berlin Heidelberg GmbH

Author

Rainer Liebmann
AEG Aktiengesellschaft
Sedanstr. 10, D-7900 Ulm, FRG

ISBN 978-3-540-16473-9   ISBN 978-3-540-39812-7 (eBook)
DOI 10.1007/978-3-540-39812-7

This work is subject to copyright. All rights are reserved, whether the whole or part of the material
is concerned, specifically those of translation, reprinting, re-use of illustrations, broadcasting,
reproduction by photocopying machine or similar means, and storage in data banks. Under
§ 54 of the German Copyright Law where copies are made for other than private use, a fee is
payable to "Verwertungsgesellschaft Wort", Munich.

© by Springer-Verlag Berlin Heidelberg 1986
Originally published by Springer-Verlag Berlin Heidelberg New York Tokyo in 1986
Softcover reprint of the hardcover 1st edition 1986

2153/3140-543210

To Gudrun Behnke,
for substantial help and encouragement

STATISTICAL MECHANICS OF PERIODIC

FRUSTRATED ISING SYSTEMS

Rainer Liebmann

Max-Planck-Institut für Festkörperforschung[*)]

Heisenbergstr. 1, D-7000 Stuttgart 80

CONTENTS

1. Introduction and survey   1

   1.1 Critical phenomena at second order phase transitions   4

   1.2 Scope of this book   6

2. One-dimensional frustrated Ising systems   7

   2.1 Periodic ANNNI-chain   7
       2.1.1 Groundstate degeneracy of the ANNNI-chain   9
       2.1.2 Periodic ANNNI-chain for T ≠ 0   11

   2.2 Decorated chains   15

   2.3 Partially frustrated chains   20
       2.3.1 Periodic frustrated chains   21
       2.3.2 Random frustrated chain   22

3. Two-dimensional frustrated Ising systems   26

   3.1 Transformations of Ising systems   26
       3.1.1 Duality transformation   29
       3.1.2 Decimation transformation   34
             a)   Decoration-iteration transformation   34
             b)   Star-triangle transformation   35

       3.1.3 Connection between different lattices   37

[*)]
   present address: AEG Aktiengesellschaft, Sedanstr. 10, D-7900 Ulm/FRG

3.2   Triangular lattice   38

    3.2.1   Estimation of the groundstate degeneracy   39

        a)   Simple lower bound   41

        b)   Pauling method   42

        c)   Systematic cluster approximation   42

    3.2.2   Partition function and exact GS entropy of the
        isotropic system   46

    3.2.3   Specific heat near the frustration points
        $(J_1 = J_2)$ 48

    3.2.4   Pair correlation function, disorder lines
        $(J_1 = J_2)$ 51

    3.2.5   Mapping to the quantum xy-chain and to the
        kinetic nn Ising chain   54

3.3   Further frustrated systems with noncrossing interactions 61

    3.3.1   Union-Jack lattice   61

    3.3.2   Villain's odd model and its generalizations   68

        a)   Groundstates and phase diagrams   68

        b)   Correlation functions   70

        c)   Periodical layered models   71

        d)   Chessboard model   74

    3.3.3   Hexagon lattice   75

    3.3.4   Pentagon lattice   76

    3.3.5   Kagomé lattice   77

    3.3.6   Connection between groundstate degeneracy and
        existence of a phase transition at $T_c = 0$ 81

3.4   Frustrated Ising systems with crossing interactions   82

    3.4.1   2d ANNNI-model   82

    3.4.2   Brick model   87

    3.4.3   Frustrated triangular lattice with nnn-interac-
        tions and magnetic field   88

        a)   Additional nnn-interactions J_2 89

        b)   Additional magnetic field H   93

        c)   Corresponding lattice gas model   97

    3.4.4   Square lattice with competing nn- and nnn-inter-
        actions: relation to vertex models   99

        a)   System without magnetic field   99

        b)   Systems with magnetic field   101

        c)   Connection to vertex models   105

4.   Three-dimensional frustrated Ising systems   109

4.1   fcc antiferromagnet   109

4.2   Fully and partially frustrated simple cubic lattice   112

4.3 AF pyrochlore model   117

4.4 ANNNI-model   122

4.5 fcc four-spin (quartet) model   127

5. Conclusion   131

    References   133

1. Introduction and Survey

The present book[*] reviews the theory of phase transitions in Ising systems[1] with competing interactions. Because of this competition no configuration of Ising spins $(s_i = \pm 1)$ can minimize the energy of all interactions simultaneously. Even in the groundstate $(T = 0)$ some interactions are 'broken', i.e. remain in the energetically unfavorable configuration.

The competition of the interactions leads, depending on their relative strength, to a multitude of commensurate, but also incommensurate phases. If the corresponding transition is of second order, different critical exponents may occur, including nonuniversal behavior. For certain ratios of the interactions the degeneracy of the groundstates becomes especially large, and the pair correlation function at $T = 0$ may decrease with a power law or exponentially. The present review tries to summarize the results obtained up to now in this fast developing field.

As the simplest example for competing interactions we consider a system of three Ising spins at the corners of a triangle (Fig. 1.1) with the Hamiltonian

$$H = - (J_1 s_2 s_3 + J_2 s_3 s_1 + J_3 s_1 s_2) \qquad (1.1)$$

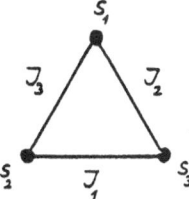

Fig. 1.1

The triangle formed by the three spins $s_i = \pm 1$ with the pair interactions J_j .

In the case of antiferromagnetic interactions for topological reasons at least one interaction is always 'broken'. Depending on the relative strength of the three interactions J_1, J_2 and J_3 a dif-

---

[*] This book is based on the habilitation thesis of the author, which has been accepted in April 1985 by the Physics Department of the University Frankfurt, Fed. Rep. Germany.

ferent groundstate degeneracy N_g occurs (Fig. 1.2):

(a) $J_1 \leq J_2 < J_3 < 0$: J_3 is weakest, therefore broken for $T = 0$;
$\underline{N_g = 2}$, (as in the ferromagnetic case),
configuration (α)

(b) $J_1 < J_2 = J_3 < 0$: either J_2 or J_3 broken for $T = 0$;
$\underline{N_g = 4}$, configurations $(\alpha) + (\beta)$

(c) $J_1 = J_2 = J_3 < 0$: either J_1 , J_2 or J_3 broken for $T = 0$;
$\underline{N_g = 6}$, configurations $(\alpha) + (\beta) + (\gamma)$.

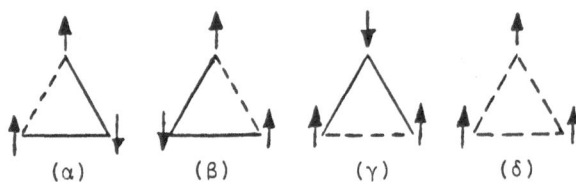

(α) (β) (γ) (δ)

Fig. 1.2   The four possible configurations of broken interactions (dashed lines)
in the triangle with antiferromagnetic interactions.

In the isotropic case (c) the degeneracy is extreme, six of the eight
possible states are groundstates. In a weak finite magnetic field H
N_g is reduced by a factor of 2 . In Section 3.2 we shall see that
in the isotropic case for H = 0 also in the infinite triangular
lattice the groundstate entropy per site remains finite and $T_c = 0$.

In the context of investigations of spin glasses[2] Toulouse[3] has coined
the term frustration for systems with competing interactions; we use
both terms equivalently. In the theoretical treatment of spin glasses
apart from frustration the second problem is the topological disorder
of the magnetic ions in the lattice, which is usually modelled as a
disorder of the nearest neigbor (nn) interaction strength between
spins on a regular lattice. But even for such models there exist almost no exact results. The question, whether the spin glas transition
showing up e.g. as a cusp in the susceptibility is a real phase transition or 'only' a dynamical nonequilibrium effect, is still not completely resolved.

This complication is one of the reasons for the interest in periodic frustrated systems, where at least for two-dimensional (2d) systems many exact results are known. In periodic frustrated systems too meta-stable states with extremely long lifetime may occur leading to spin glas-like behavior. Therefore, this periodic systems are not only in-teresting in themselves, but also well suited to test approximation methods for treating spin glasses.

Experimentally, systems with competing interactions are of great sig-nificance. For example, in many magnetic substances the interactions are often not limited to nearest neighbors (nn) . If nn interactions alone lead already to frustration, as for antiferromagnetic interac-tions in lattices consisting of joint triangles (2d) or tetrahedra (3d) , additional interactions become very important. Even if they are weak they can reduce the groundstate degeneracy significantly. Weak additional interactions may also lead to a crossover from lower- to higher-dimensional behavior with decreasing temperature.

Ising systems are used to describe magnetic substances, where e.g. be-cause of local crystal fields the magnetic moment (spin) of the single ions can only be oriented in two opposite directions (spin dimension $n = 1$). In this respect they differ from xy- and Heisenberg systems, where the spins can move on a circle $(n = 2)$ or on a sphere $(n = 3)$ respectively. Although we only regard Ising systems here, for conven-ient interactions (often connected to enhanced groundstate degeneracy) we will find also phase transitions e.g. of xy-type.

Apart from describing magnetic substances, Ising systems are also re-levant for very different physical systems. The absorption of (sub-) monolayers of atoms on a crystalline surface can be mapped on a two-dimensional Ising system, if absorption takes place only on a discrete periodic lattice of absorption sites determined by the substrate. Spins $s_i = +1$ correspond to an absorption site occupied by an atom, $s_i = -1$ to a vacancy. For the proper description of phase transitions in these monolayers (e.g. from commensurate to incommensurate) apart from nn interactions one needs competing interactions between more distant sites.

Substitution alloys $A_x B_{1-x}$ consisting of two types of atoms A, B can also be described as Ising systems, if the atoms cannot go on interstitial sites. Then $s_i = 1$ corresponds to an A atom at site i , and $s_i = -1$ to a B atom.

As a last example for the many applications of Ising systems to experiments, we mention the structural order-disorder transitions, e.g. for the position of the protons in ice, or the transitions in ferroelectrics described originally by different vertex models. For instance the low-temperature degeneracy of ice can be mapped on the groundstate degeneracy of a frustrated Ising system, where the spin orientation $s_i = \pm 1$ corresponds to the two allowed positions of the proton between the two adjacent O^{2-}-ions.

After this discussion of the connection of frustrated Ising systems to very different physical fields, in Section 1.1 we give a very short summary of the critical properties at second order phase transitions, followed in Section 1.2 by a detailed survey of the present review.

1.1 Critical Phenomena at Second Order Phase Transitions

Many physical systems show a sudden change of their properties at a well defined critical temperature T_c . Often such phase transitions are accompanied by singularities in thermodynamic quantities as the free energy, the specific heat and the susceptibility.

If it is possible to introduce an order parameter, which is finite for $T < T_c$, but vanishes for $T > T_c$ (e.g. the magnetisation M in ferromagnetic systems), phase transitions of second order are characterized by a continuous, but nonanalytic behavior of the order parameter at T_c . Defining the reduced temperature $t = |T-T_c|/T_c$, in the terminology of magnetic systems one finds

$$M(t) \; \propto \; t^\beta \qquad (T < T_c) \qquad . \qquad\qquad (1.2a)$$

Here β is the critical exponent describing the asymptotic behavior close to the phase transition. For Ising systems the order parameter M is a scalar $(n = 1)$. Other properties show analogous behavior at T_c . For the specific heat c , the susceptibility χ , the correlation length ξ and the asymptotic pair correlation function $G(r)$ for $r \to \infty$ further critical exponents can be defined:

$$c(t) \; \propto \; t^{-\alpha} \; , \qquad \chi(t) \; \propto \; t^{-\gamma} \; , \qquad \xi(t) \; \propto \; t^{-\nu} \; : \; H = 0$$

$$m(H) \; \propto \; H^{1/\delta} \; , \qquad G(r) \; \propto \; r^{2-d-\eta} \; : \; t = 0 \qquad . \qquad (1.2b)$$

Experiments have shown, that whole 'universality classes' of very different substances have identical critical exponents. This could not be understood until scaling hypothesis and renormalization group theory[4] were developed. Thus in ferromagnetic systems with ferromagnetic nearest neighbor (nn) pair interactions the critical exponents depend only on the spin dimension n and the lattice dimension d , but not on the strength of the interactions, and also not (!) on the specific lattice type. For example, the ferromagnetic Ising systems on the triangular and square lattice have the same critical exponents.

For d = 2 and d = 3 the normal Ising exponents are[5]:

Table 1.1 Normal Ising exponents for d = 2 and d = 3 .

d	α	β	γ	δ	η	ν
2	O(log)*	1/8*	7/4*	15.04±0.07	1/4*	1*
3	0.013±0.01	$0.312^{+0.002}_{-0.005}$	1.250±0.002	5.0±0.05	–	$0.638^{+0.002}_{-0.001}$

In Table 1.1 stars mark exact results for d = 2 .

If frustration occurs in an Ising system, the order parameter may have more than one component, as mentioned above. General considerations in this connection have been worked out e.g. by Mukamel and Krinsky[6] and Alexander and Pincus[7]. As the dimension of the order parameter in frustrated systems depends on the detailed lattice structure and not only on d , one can expect a much richer variety of critical behavior as compared to the nonfrustrated case.

Until now we have regarded only two-spin interactions. Inclusion of a magnetic field (one-spin interaction) and multi-spin interactions yields more general models with different universality classes. In this review we will consider almost exclusively pair-interactions, but a few results for four-spin interactions are added.

1.2 Scope of This Book

The following three chapters (2 to 4) of this book are arranged according to the lattice dimension d of the various periodic frustrated
systems (d = 1 to 3) . 1d Ising systems with short range interactions
are usually simple to solve, and for 2d systems without crossing
interactions e.g. transfer matrix methods yield exact solutions. In
contrast to this for 2 d systems with crossing interactions and in
general for 3d systems only approximate methods are available, such
as high- and low-temperature expansion, mean field- (MF-) approximation, cluster-approximations, renormalization group (RG) methods and
Monte Carlo (MC) calculations.

Chapter 2 deals in Section 2.1 with the properties of the ANNNI-chain
for T = 0 and T ╪ 0 . Section 2.2 describes decorated and Section
2.3 partially frustrated Ising chains.

Chapter 3 is by far the longest one, as for d = 2 a large number of
exact results is known. After general transformations of Ising systems (in 3.1) the antiferromagnetic (AF) Ising model on the triangular lattice is treated extensively. Besides the groundstate (GS) degeneracy, the specific heat, the asymptotic behavior of the pair correlation function, the related disorder line, and the mapping on the
quantum-XY- and the kinetic Ising chain are discussed. In Section 3.3
a couple of further frustrated 2d Ising systems without crossing
interactions are considered, which differ for instance in GS entropy and the decay of the correlation function and, therefore, belong
to different universality classes.

Finally Section 3.4 summarizes the properties of three systems with
crossing interactions. Especially, the first of them, the ANNNI-model, is compared to the exactly solvable brick model in Section 3.4.2.

Chapter 4 reviews the 3d systems, which depending on the specific
lattice type differ very much and belong to different universality
classes.

Chapter 5 finally contains a short summary of this review on periodic
frustrated Ising systems.

2. One-Dimensional Frustrated Ising Systems

To explain the expression frustration, in the introduction the triangular Ising cluster with equal antiferromagnetic interactions was mentioned as the simplest example. Without a magnetic field six, with a
field three degenerate groundstates occur. In this chapter we now consider one-dimensional Ising systems, obtained by different ways of
connecting frustrated triangles. We treat these one-dimensional systems because, in spite of the mathematical simplicity to solve them,
they show already many similarities to the higher-dimensional frustrated systems discussed later. Of course real long range order (LRO)
can occur in one-dimensional systems with short range interactions
only for $T = 0$, as $d = 1$ is the lower critical dimension of such
systems. In Fig. 2.1 several possibilities are shown to form a chain
from triangles, which we are going to discuss more closely now.

2.1 Periodic ANNNI-Chain

In Fig. 2.1a the triangles have common edges. By redrawing it
(Fig. 2.1a') it is easily seen, that the interactions J_1 and J_2 are
the nn and nnn interactions in a linear chain. This linear chain is
the one-dimensional version of the ANNNI-model (axial next nearest
neighbor Ising-model)[8]; its two- and three-dimensional versions with
characteristic different properties will be discussed later.

The Hamiltonian for $d = 1$ is:

$$H = - J_1 \sum_i \sigma_i \sigma_{i+1} - J_2 \sum_i \sigma_i \sigma_{i+2} \qquad (2.1)$$

with antiferromangetic (AF) $J_2 < 0$; the sign of J_1 does not matter without a magnetic field, it only determines (see Fig. 2.1a) the
relative orientation of one half-chain to the other one.

Substituting[9] $\sigma_i \sigma_{i+1} = s_i$ the Hamiltonian (2.1) is mapped into

$$H = - B \sum_i s_i - J \sum_i s_i s_{i+1} \quad , \quad \text{with } B = J_1 , J = J_2 ,$$

$$(2.2)$$

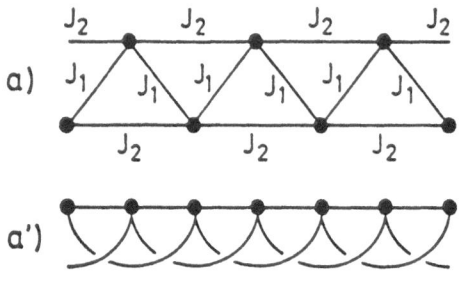

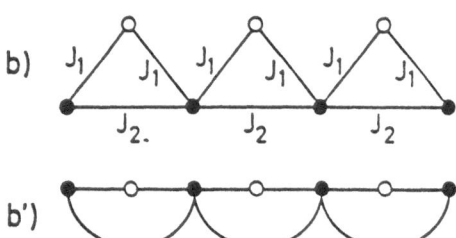

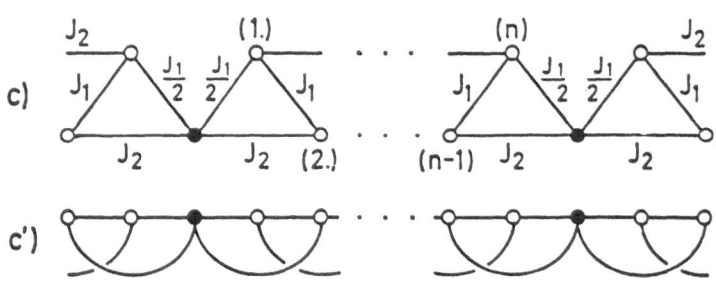

Fig. 2.1 Frustrated chains formed by connecting frustrated triangles in different ways:

(a) Periodic ANNNI chain with nn and nnn interactions J_1 and J_2 ;

(b) simple decorated chain; ● 'knot spins', ○ 'decorated spins';

(c) n-decorated chains (= Mock ANNNI chain).

In (a') to (c') the chains are drawn strictly one-dimensional.

i.e. the chain with nn- and nnn-interactions is equivalent to the chain with antiferromagnetic nn-interactions in a homogeneous external field, the sign of which of course is irrelevant. We will use this form of the Hamiltonian later to calculate the pair correlation function.

2.1.1 Groundstate Degeneracy of the ANNNI-chain

Let us consider first the groundstate of (2.1). For $J_1 > 0$ it depends on the ratio $\alpha = -J_1/J_2$. For $\alpha < 1/2$ J_1 dominates and there are only two simple ferromagnetic groundstates. For $\alpha > 1/2$ J_2 dominates, and there are now four groundstates with alternating two spins up, two spins down, which we call <2>-groundstates (e.g. ref. 10).

Whereas for $\alpha \lesssim 1/2$ only groundstates with long range order (LRO) occur, for $\alpha = 1/2$ there is no LRO . For $\alpha = 1/2$ the chain is "completely frustrated", the point $(\alpha = 1/2, T = 0)$ in phase space is called a frustration point[11]. There exists a large number of ground-states, for which the only condition now is the requirement of at least one parallel nn spin for each spin.

The number N_o of the groundstates (GS) and the pair correlation functions for $\alpha = 1/2$ averaged over these GS can be calculated from simple recursion formulas. Defining the number of GS of a chain of j spins, ending with orientations ++, +-, -+, -- , with N_1^j, N_2^j, N_3^j, N_4^j , one obtains immediately the recursion formulas:

$$N_1^{j+1} = N_1^j + N_3^j \quad ,$$

$$N_2^{j+1} = N_1^j \quad ,$$

$$(2.3)$$

$$N_3^{j+1} = N_4^j \quad ,$$

$$N_4^{j+1} = N_4^j + N_2^j \quad .$$

Combining the GS ending with a (+)- or (-)-spin to $N_p^j = N_1^j + N_3^j$ and $N_m^j = N_2^j + N_4^j$ respectively, from (2.3) follows:

$$N_p^{j+1} = N_p^j + N_m^{j-1} \quad ,$$

$$(2.4)$$

$$N_m^{j+1} = N_m^j + N_p^{j-1} \quad .$$

For the total number of GS , $N_o^j = N_p^j + N_m^j$ and for the difference $N_{pm}^j = N_p^j - N_m^j$ one gets the recursion formulas:

$$N_o^{j+1} = N_o^j + N_o^{j-1} \qquad , \qquad (2.5a)$$

$$N_{pm}^{j+1} = - N_{pm}^{j-2} \qquad . \qquad (2.5b)$$

As (2.5a) is the well known definiton of a Fibonacci-series, in the limit $j \to \infty$ one obtains (independent of the first two values N_o^2 and N_o^3) :

$$N_o^j \propto q^j \qquad \text{with} \qquad q = \frac{\sqrt{5}+1}{2} \qquad . \qquad (2.6)$$

The resulting GS entropy per lattice site S_o for $\alpha = 1/2$ is:

$$S_o = \lim_{j \to \infty} \frac{1}{j} \ln N_o^j = \ln q \approx 0.4812 \qquad , \qquad (2.7)$$

whereas for $\alpha \neq 1/2$ the GS entropy vanishes. Thus $S(T=0)$ is not a continuous function in α for $\alpha = 1/2$. For the pair correlation function one obtains

$$G(r) = \langle \sigma_1 \sigma_{1+r} \rangle_o \propto \frac{N_p^{r+1} - N_m^{r+1}}{N_p^{r+1} + N_m^{r+1}} \propto \frac{N_{pm}^r}{q^r} = e^{-r \ln q} \cdot N_{pm}^r \qquad . \qquad (2.8)$$

From (2.5b) N_{pm}^j is periodic in j with period 6 .

For chains beginning with (++)-spins or (+-)-spins, for N_{pm}^r one obtains the simple periodic series (period 6) :

$$\left. \begin{array}{l} \text{(++)-start:} \quad 1 \ 0 \ \bar{1} \ \bar{1} \ 0 \ 1 \ \dots \\[2mm] \text{(+-)-start:} \quad \bar{1} \ \bar{1} \ 0 \ 1 \ 1 \ 0 \ \dots \end{array} \right\} \text{for} \quad r = 2, \ 3, \ 4, \ 5, \ 6, \ 7, \ \dots \qquad . \qquad (2.9)$$

In the middle of a chain both start configurations are not equally probable, the correct linear combination is obtained by the symmetry requirement $\langle \sigma_n \sigma_{n+r} \rangle = \langle \sigma_n \sigma_{n-r} \rangle$ for $1 \ll n-r < n + r \ll j$:

$$N_{pm}^{r} \quad \propto \quad 2 \ N_{pm}^{(++)r} + N_{pm}^{(+-)r} \quad = \quad (2), \ 1, \ \overline{1}, \ \overline{2}, \ \overline{1}, \ 1, \ 2, \ \ldots$$

$$(2.10)$$

$$\text{for} \quad r \ = \ (1), \ 2, \ 3, \ 4, \ 5, \ 6, \ 7, \ \ldots \quad .$$

Finally from (2.6), (2.8) and (2.10) the asymptotic pair correlation function $G(r)$ in the infinite chain for $\alpha = 1/2$ is

$$\lim_{r \to \infty} G(r) \quad \propto \quad e^{-r/\xi_O} \cos\left(\frac{\pi}{3} \cdot r\right)$$

$$(2.11)$$

with $\qquad \xi_O^{-1} \ = \ \ln\left(\frac{\sqrt{5}+1}{2}\right) \quad .$

This exponential decay multiplied by an oscillation of period 6 at the frustration point $\alpha = 1/2$ we will find again for finite temperature in the limit $T \to 0$. Here we only note, that for higher-dimensional frustrated systems the groundstate averages do not necessarily agree with the limit $T \to 0$.

2.1.2 Periodic ANNNI-chain for $T \neq 0$

We now want to calculate the pair correlation function for finite temperature using the transfer matrix method.

For the purpose it is convenient to take the Hamiltonian in the transformed form (2.2). As is shown e.g. by Hornreich, Liebmann, Schuster and Selke[12,13] the pair correlation function of a chain with j spins (with periodic boundary conditions) can be written by products of transfer matrices:

$$<\sigma_1 \sigma_{1+r}> \ = \ \text{Tr} \ (\tilde{\tau}^r \tau^{j-r})/\text{Tr}(\tau^j) \quad , \quad (2.12)$$

with two (2×2) transfer matrices $(s_i = \pm 1)$:

$$\tau(s_i, s_{i+1}) \ = \ \exp\left\{K_2 s_i s_{i+1} + K_1 (s_i + s_{i+1})/2\right\}$$

and

$$\tilde{\tau}(s_i, s_{i+1}) \ = \ s_i \ \tau(s_i, s_{i+1}) \ s_{i+1} \quad , \quad (2.13)$$

where $K_1 = \beta J_1$, $K_2 = \beta J_2$ and $\beta = 1/k_B T$.

Diagonalizing τ and $\tilde{\tau}$ and using the respective transformation matrices one obtains in the limit $j \to \infty$ [18]:

$$
G(\underline{r}) = \begin{cases}
a \, e^{-r\ln(\lambda_1/\tilde{\lambda}_1)} + (1-a) \, e^{-r\ln(\lambda_1/\tilde{\lambda}_2)} & \text{for } \tilde{\lambda}_1 \neq \tilde{\lambda}_2 \\[2ex]
e^{-r\ln(\lambda_1/\tilde{\lambda})} \, (1+br/\tilde{\lambda}) & \text{for } \tilde{\lambda}_1 = \tilde{\lambda}_2 = \tilde{\lambda}
\end{cases}
$$

$$(2.14)$$

with the eigenvalues of τ and $\tilde{\tau}$:

$$
\lambda_{1,2} = e^{K_2} \cosh K_1 \pm \left(e^{2K_2} \sinh^2 K_1 + e^{-2K_2} \right)^{1/2} ,
$$

$$
\tilde{\lambda}_{1,2} = e^{K_2} \sinh K_1 \pm \left(e^{2K_2} \cosh^2 K_1 - e^{-2K_2} \right)^{1/2} . \quad (2.15)
$$

It is interesting to note the two regions in phase space $(\alpha = J_2/J_1$, $K_1^{-1} = k_B T/J_1)$ according to (2.14) with different behavior of $G(r)$.

For $\cosh K_1 > e^{-2K_2}$ both $\tilde{\lambda}_1$ and $\tilde{\lambda}_2$ are real, and one obtains a simple exponential decay of the pair correlation function with the correlation length

$$
\xi = \left(\ln (\lambda_1/\tilde{\lambda}_1) \right)^{-1} , \quad\quad (2.16)
$$

whereas for $\cosh K_1 < e^{-2K_2}$ $\tilde{\lambda}_{1,2}$ are complex, and the exponential decay with

$$
\xi = (\ln |\lambda_1/\tilde{\lambda}_1|)^{-1} \quad\quad (2.16')
$$

is multiplied by an oscillation with wavevector

$$
q = \arctan (\operatorname{Im} \tilde{\lambda}_1/\operatorname{Re} \tilde{\lambda}_1) . \quad\quad (2.17)
$$

The boundary between these two regions

$$
\cos K_1 = e^{-2K_2} \quad\quad (2.18)
$$

in ref. 12 we called 'Lifshitz condition'. Subsequently it turned out, that Stephenson[13] obtained the same results earlier. He called the boundary between oscillating and nonoscillating decay of the pair correlation function a 'disorder line'. In addition he distinguished two kinds of them: In disorder lines of the

a) <u>first kind</u>   the wavevector q of the oscillation varies continuously e.g. with temperature, as in our example;

b) <u>second kind</u> q is temperature independent. This is true for example for the chain shown in Fig. 2.1c.

Figure 2.2[14] shows the dependence of q , Eq. (2.17), on α and the temperature K_1^{-1} . In the region on the left side below the full line no oscillations occur $(q \equiv 0)$, whereas to the right of this line q varies continuously; the dashed lines correspond to fixed values of q .

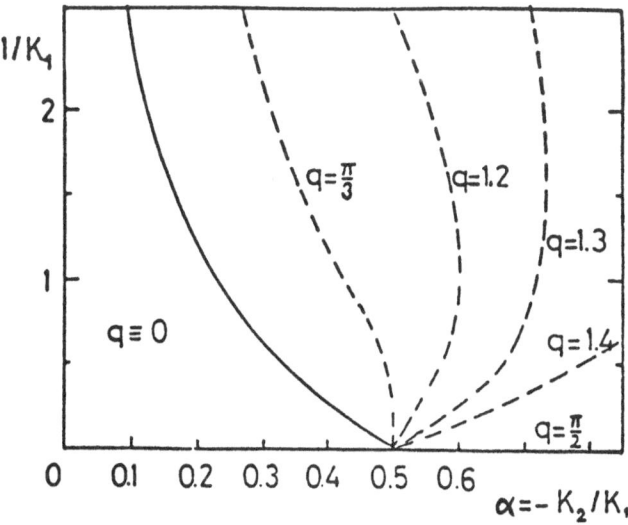

<u>Fig. 2.2</u> Lines of constant wavevector in the K_1^{-1}-α-plane; on the left side below the (fully drawn) disorder line $q \equiv 0$.

The frustration point P_F ($\alpha = 0.5$, $T = 0$) obviously is a singular point, because the limit value of q at P_F depends on the direction, from which P_F is approached.

Figure 2.3 shows the behavior of ξ for $\alpha \le 1/2$ [13]. As the ground-state is ferromagnetic for $\alpha < 1/2$, $\lim \xi^{-1} = 0$. Precisely at the disorder line where the decay of the correlation function changes from monotonous to oscillating, Eq. (2.18), one finds maxima in ξ^{-1} , e.g. minima in ξ , justifiing the name 'disorder line'.

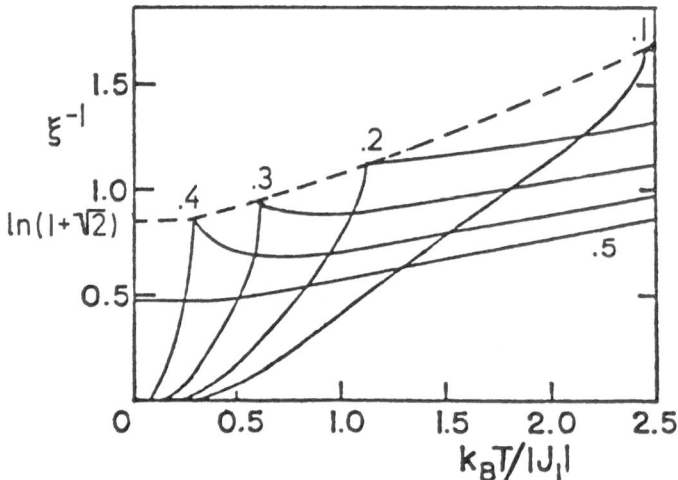

Fig. 2.3   Temperature dependence of the inverse correlation length ξ^{-1} (T)   for fixed values of α . ξ^{-1} shows cusps along the disorder line ($\alpha < 0.5$, compare with Fig. 2.2).

The limit of ξ^{-1} approaching P_F along the disorder line is

$$\xi_d^{-1} \;=\; \ln\,(\sqrt{2}+1) \quad . \tag{2.19}$$

In contrast to this for $\alpha = 1/2$ $\xi^{-1}(T)$ does not show any spike, and also does not vanish for $T \to 0$, but approaches a finite value:

$$\xi_o^{-1} \;=\; \ln\left(\frac{\sqrt{5}+1}{2}\right) \quad , \tag{2.20}$$

with $q \to q_O = \frac{\pi}{3}$, in full agreement with the result (2.11) obtained
from averaging over the groundstates.

Besides the pair correlation function also other thermodynamic quantities can be determined by the transfer matrix method, e.g. the
internal energy U , the entropy S and the specific heat c (each
per lattice site):

$$U = - \ln \lambda_1/\beta \quad , \tag{2.21}$$

$$S/k_B = \ln \lambda_1 - \beta \, (\partial \ln \lambda_1/\partial \beta) \quad , \tag{2.22}$$

$$c = \beta^2 \, (\partial^2 \ln \lambda_1/\partial \beta^2) \quad , \tag{2.23}$$

with λ_1 from (2.15). Figure 2.4a,b[15] clearly shows the asymmetric
maximum of the entropy (2.22) near the value $\alpha = 0.5$, which for
$T \to 0$ (or $K_1 \to \infty$) becomes increasingly sharp and reproduces exactly
the singular value $S_O = \ln \, (\sqrt{5}+1)/2) \simeq 0.481$, Eq. (2.7), at the
frustration point $(\alpha = 0.5, T = 0)$.

In addition we note the completely monotonous behavior of the entropy
S (as well as U and c) in the vicinity of the disorder line, as these
quantities are independent of $\tilde{\lambda}_{1,2}$.

2.2 Decorated Chains

After the relatively detailed discussion of the periodic ANNNI-chain
of Fig. 2.1a we continue with a shorter look at the decorated chains
of Fig. 2.1b,c.

The simple decorated chain (Fig. 2.1b) has also been treated by
Stephenson[13]. By dedecoration[16], that is by summation over the decorated spins coupled only to nn-spins one obtains an effective
temperature dependent nn-interaction of the remaining 'knot spins'

$$K_1^{eff} = K_2 + 1/2 \ln \cosh 2K_1 \quad . \tag{2.24}$$

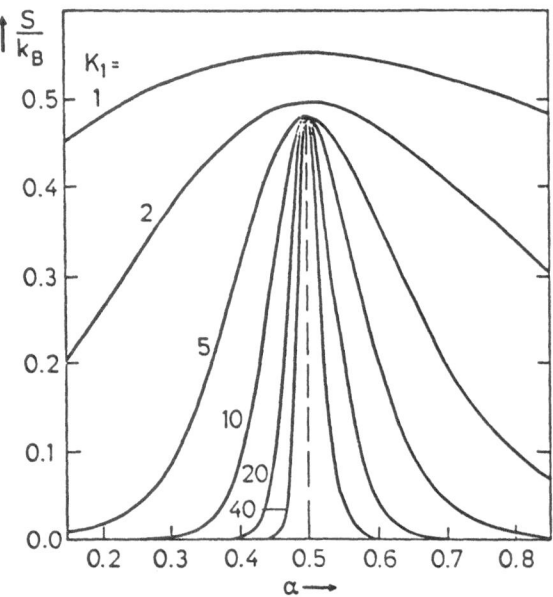

Fig. 2.4a Entropy S as function of α for fixed values of the inverse temperature K_1 .

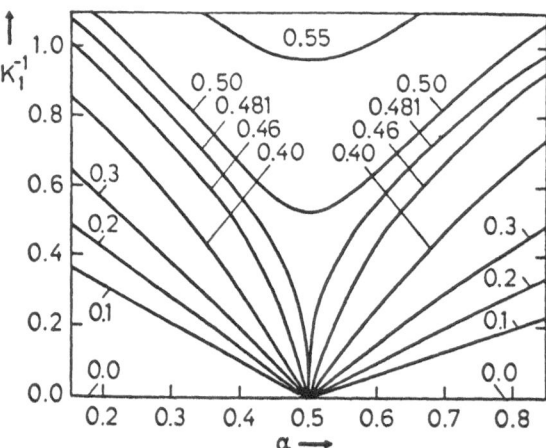

Fig. 2.4b Lines of constant entropy S in the T-α-plane. At the frustration
point S is not analytic.

This effective interaction vanishes for

$$\cosh 2K_1 = e^{-2K_2} \; . \tag{2.25}$$

The condition (2.25) defines a disorder line analogous to the ANNNI-chain, Eq. (2.18), but now for the interval $0 \le \alpha \le 1$. As opposed to that case here K_1^{eff} just changes sign along the disorder line, and below the temperature T_d it is antiferromagnetic $(q = \pi/2)$, in both cases with exponential decay. The frustration point, where even for $T = 0$ no LRO exists, occurs at $(\alpha = 1 , T = 0)$.

As K_1^{eff} vanishes for $T = T_d$, so does ξ . This kind of disorder line, where q switches from 0 to (the temperature independent value) $\pi/2$, has been called a disorder line of the second kind by Stephenson[13]. The behavior of $\xi_{eff}(T)$ from ref. 13 is shown in Fig. 2.5.

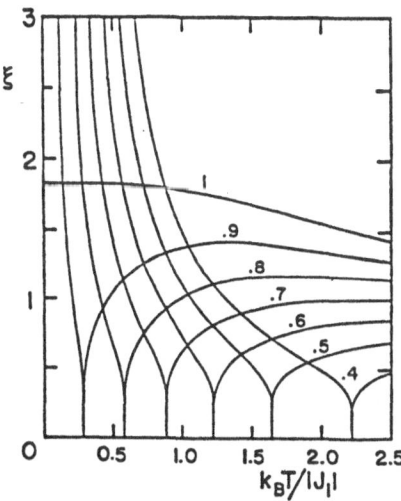

Fig. 2.5 Temperature dependence of the correlation length ξ of the simple decorated chain for fixed values of α (indicated). Along the disorder line $\xi = 0$. (Ref. 13)

After this consideration of the simple decorated chain (Fig. 2.1b) we can now easily treat the generalized chain of Fig. 2.1c. This model has been investigated in connection with higher-dimensional uniaxial ANNNI-models[17]. Inbetween the knot spins (•) it now contains n instead of 1 decorated spin (o), and the nn-interaction of the knot spins to their nn decorated spins is reduced to $J_1/2$, to shift the frustration point to the ANNNI-value ($\alpha = 0.5$, $T = 0$). Using transfer matrices the decorated spins can be summed out, leaving again an effective nn-interaction $K_1^{eff}(K_1, K_2)$.

Because of the n decorated spins K_1^{eff} now can change sign repeatedly. As has been shown[17], K_1^{eff} vanishes $[(n+2)/2]$ times, where $[x]$ is the integer part of x.

For given K_1, the zeros are given explicitly by

$$\alpha = \frac{1}{2}\left[1 + \frac{1}{K_1}\ln\left(\frac{1}{2}\sqrt{(1+e^{-2K_1})^2 + (1-e^{-2K_1})^2\tanh^2\left(\frac{m}{n}\pi\right)}\right)\right] ,$$

$$(2.26)$$

with $m = 0, \ldots, [n/2]$, and are shown as full lines in Fig. 2.6[18].

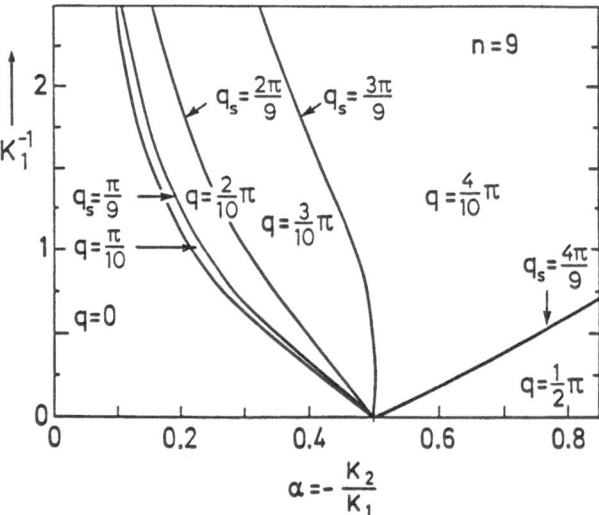

Fig. 2.6 Disorder lines of the second kind of the (n=9)-decorated chain (from Eq. (2.26)). In the regions between these lines the wavevector is constant: $q = \frac{m}{10}\pi$, $0 \le m \le 5$.

For $T \propto K_1^{-1} \to 0$ asymptotically these are straight lines through the frustration point with slope

$$c = \frac{-2}{\ln\left(2 \cos\left(\frac{m}{n}\pi\right)\right)} \ . \tag{2.27}$$

As a straightforward generalization of the simple decorated chain the chain with n decoration spins thus shows $\left[\frac{n+2}{2}\right]$ disorder lines of the second kind, where the wavevector q switches from one fixed value $q_m^{(-)}$ to the next one $q_m^{(+)}$:

$$q_m^{(-)} = \frac{m}{n+1} \cdot \frac{\pi}{\alpha} \ ,$$

$$\tag{2.28}$$

$$q_m^{(+)} = \frac{m+1}{n+1} \cdot \frac{\pi}{\alpha} \ .$$

The first, n-independent disorder line ($m = 0$ in (2.23)) forming the boundary of the nonoscillating region ($q \equiv 0$) is identical to the single disorder line (2.18) of the periodic ANNNI-chain. The other $[n/2]$ ones occur in addition and are n-dependent. With increasing n in the n-decorated chain an increasing number of regions with fixed q_m occur, separated by disorder lines lying dense for $n \to \infty$. Thus one recovers the continuous variation of q (Eq. (2.17)) beyond the boundary (2.18).

The connection between the periodic ANNNI- and the n-decorated chain can be described even more precisely. Inserting (2.28) in (2.26) one obtains

$$\tilde{q}_m = \text{arc tan}\left(\frac{\text{Im } \tilde{\lambda}_1}{\text{Re } \tilde{\lambda}_1}\right) \ , \quad \text{with} \quad \tilde{q}_m = \pi\frac{m}{n} \ , \tag{2.29}$$

in analogy to (2.17), with the sole difference that because of the structure of the n-decorated chain $\tilde{q}_m$ is discrete, whereas is varies continuously in the periodic ANNNI-chain.

The disorder lines of the n-decorated chain are at the same time lines of constant wavevector in the periodic chain, with

$$q_m^{(-)} \leq q_{\text{period.}} \leq q_m^{(+)} \ . \tag{2.30}$$

The continuous respective stepwise dependence of the wavevector on α, $q = q(\alpha)$ for the periodic and the (n=9)-decorated chain is demonstrated in Fig. 2.7 for $K_1^{-1} = 0.5$.

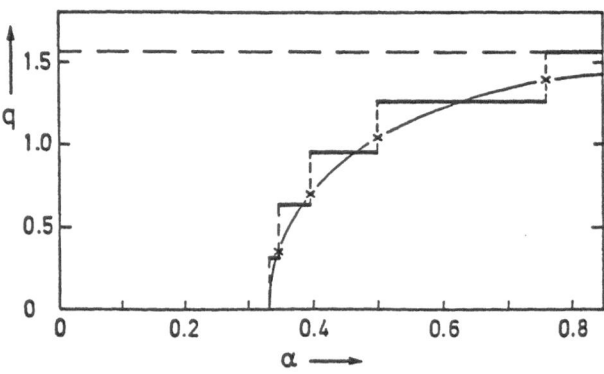

Fig. 2.7 Comparison of the α dependence of the wavevector q for the periodic ANNNI-chain (continuous) and the (n=9)-decorated chain for $T/J_1 = 0.5$.

2.3 Partially Frustrated Chains

We now turn to the last one-dimensional examples, the partially frustrated chains and distinguish between periodic and random frustrated chains. The motivation for considering the periodic case is its similarity to higher-dimensional fully frustrated systems as for example the three-dimensional pyrochlore- (or B-spinell-) lattice discussed in Chapter 4. The random case is interesting as a one-dimensional version of spin glasses, because here one can treat the disorder still analytically.

2.3.1 Periodic Frustrated Chains

Doman and Williams[19] have considered a periodic Ising chain in a homogeneous magnetic field B , where one ferromagnetic (F) and three antiferromagnetic (AF) nn-interactions are repeated periodically (here called 1-3 chain):

$$H = - \sum_i J_i \ s_i \ s_{i+1} - B \sum_i s_i \qquad ; \qquad (2.31a)$$

with

$$J_{4i} = J \quad , \quad J_{4i+1} = J_{4i+2} = J_{4i+3} = -J \ ; \ J > 0 \quad .$$

$$(2.31b)$$

Using the transformation $s_i = \sigma_i \ \sigma_{i+1}$ (compare Sec. 2.1) this 1-3 chain is equivalent to a chain with nn-interactions $J_{(1)} = B$ and nn-interactions $J_{(2)_i} = J_i$:

$$H = - \sum_i J_{(2)_i} \ \sigma_i \ \sigma_{i+2} - J_{(1)} \sum_i \sigma_i \ \sigma_{i+1} \quad . \qquad (2.32)$$

both versions of the 1-3 chain are shown in Fig. 2.8.

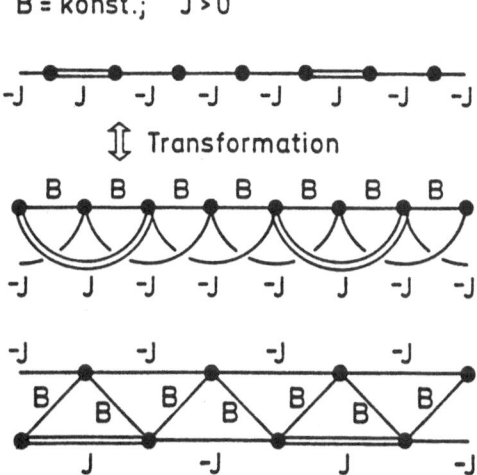

$$B = konst.; \quad J > 0$$

Fig. 2.8 1-3 chain of Doman and Williams.
═══ ferromagnetic interactions, ─── antiferromagnetic interactions.

With the transfer matrix method already mentioned, Doman and Williams[19] have determined the energy U, the entropy S, the specific heat c and the magnetisation M. As examples in Figs. 2.9 und 2.10 the entropy and the magnetisation are reproduced as functions of B/J. From these figures apparently two critical ratios $(B/J)_{c_1} = 1$ and $(B/J)_{c_2} = 2$ exist where the groundstate configuration changes.

For $(B/J) < 1$ the nn-interactions J_i determine the groundstate, with $M_o = 0$ and $S_o = 0$. For $1 < (B/J) < 2$ all pairs of F interacting spins are oriented parallel to the magnetic field, the antiparallel pairs of spins inbetween have two allowed configurations. This yields $M_o = 1/2$ and $S_o = 1/4 \ln 2 \approx 0.1733$. For $(B/J) > 2$ all spins are parallel to the field, $M_o = 1$, $S_o = 0$. Just at the critical values of the ratio B/J the groundstate entropy is especially large, like in the periodic ANNI-chain, and M_o exhibits values inbetween the neighboring 'phases'[19]:

$$S_o^{C_1} = \frac{1}{4} \ln (\sqrt{2}+1) \approx 0.2203 \quad ,$$

$$M_o^{C_1} = \frac{\sqrt{2}}{4} \approx 0.3536 \quad ,$$

$$(2.33)$$

$$S_o^{C_2} = \frac{1}{4} \ln 3 \approx 0.2747 \quad ,$$

$$M_o^{C_2} = \frac{2}{3} \quad .$$

For finite temperature this discontinuous behavior is again washed out to asymmetric continuous curves.

2.3.2 Random Frustrated Chain

The Ising chain in a homogeneous magnetic field with random ferro- and antiferromagnetic nn-interactions of concentration $(1-x)$ and x has been investigated by a number of authors[19-25].

The essential additional problem compared to the periodic frustrated chains already discussed is the necessity to average over all bond configurations for given concentration x. Whereas the pure antifer-

23

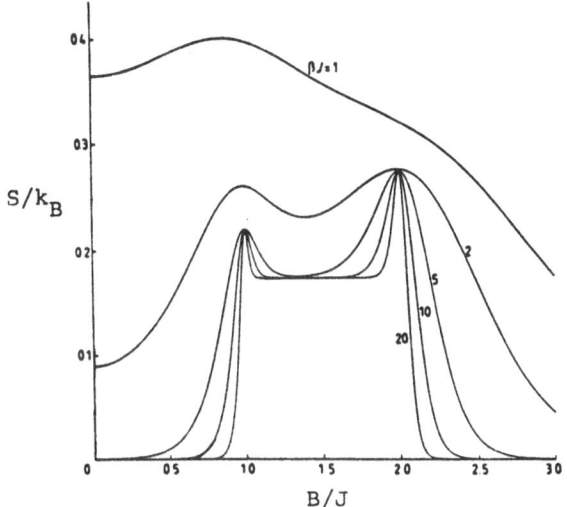

Fig. 2.9   Entropy S of the 1-3 chain as function of B/J for fixed values of
the inverse temperature J/T (ref. 19). Two critical magnetic fields
occur.

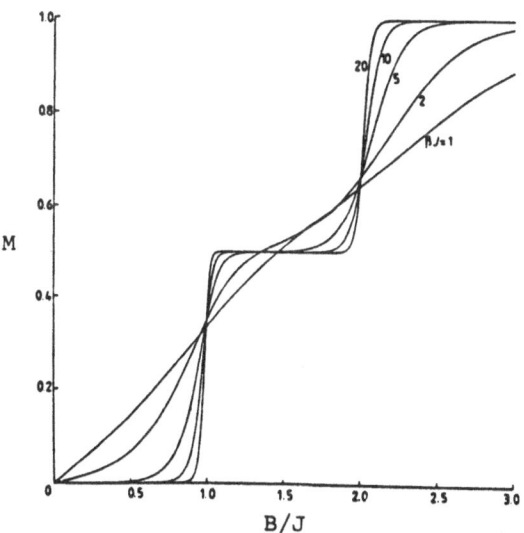

Fig. 2.10   Magnetisation M of the 1-3 chain for fixed values of the inverse
temperature J/T (ref. 19).

romagnetic chain in a magnetic field ($\hat{=}$ periodic ANNNI-chain) exhibits only one critical magnetic field $(B_c = 2J)$, and the 1-3 chain discussed above exhibits two ones, where the groundstate magnetisation shows steps, in the random bond chain there are steps in the groundstate magnetisation for all $B_c^{(r)} = 2J/r$, $r = 1, 2, 3, \ldots, \infty$. They can be traced back to flipping spinblocks, also called superspins[20], e.g. of spin clusters with r more Ising spins oriented in one direction than in the opposite one.

After previous work, for instance by Matsubara[21], Landau and Blume[22] and Puma et al.[25], Derrida et al.[24] in 1978 obtained explicit analytical expressions for the groundstate properties.

Figure 2.11 shows their result for the groundstate entropy S_O for $x = 0.5$. The series of spikes at the critical fields $B_c^{(r)} = 2J/r$ is evident, inbetween S_O is constant.

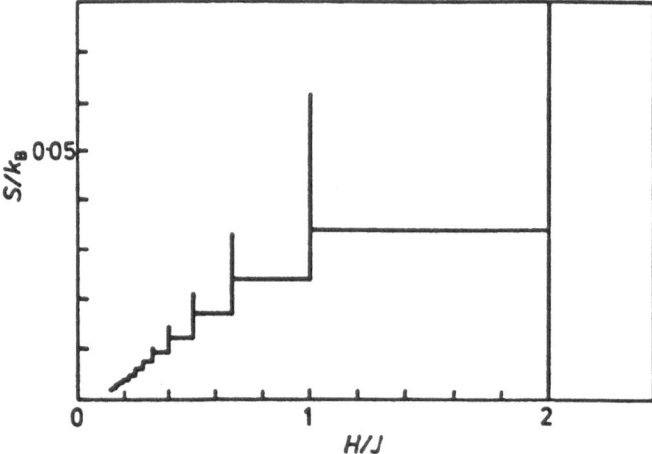

Fig. 2.11 Entropy S of the random frustrated chain as function of H/J for x = 0.5 and T = 0 ; the value of S for H/J = 2 is 0.143 , beyond the drawing (ref. 24).

The analytical expressions for S_O, at and between the critical fields are given in Derrida et al.[24].

In comparison to Fig. 2.9 here in Fig. 2.11 the continuous curves for finite temperature are not given.

Figure 2.12 shows the steps of the magnetisation M_o for $x = 0.2$, 0.5 and 0.8 [24]. Note the monotonous decrease of M_o for $B < B_{c_1} = 2J$ with increasing concentration x of the antiferromagnetic interactions. For $x \to 1$ (e.g. in the pure AF chain) only the first step at $B_{c_1} = 2J$ survives, agreeing of course with the results for the ANNNI-chain $(B \to J_1, J \to |J_2|$, thus $B = 2J \to J_1 = -2J_2$ or $\alpha = 0.5)$.

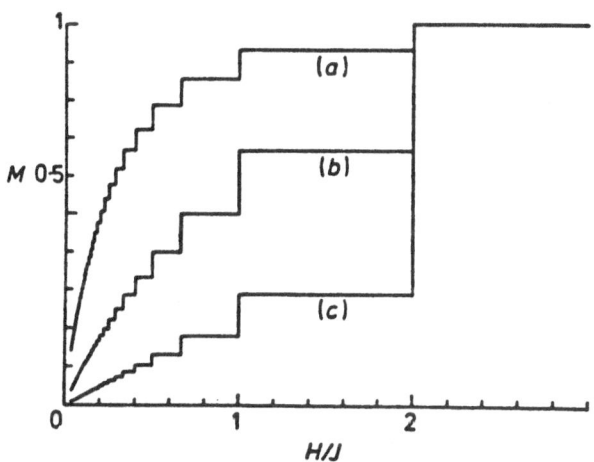

<u>Fig. 2.12</u> Magnetisation as function of H/J for three concentrations x of antiferromagnetic interactions: (a) x = 0.2 , (b) x = 0.5, (c) x = 0.8 (ref. 24).

At first Puma and Fernandez[25] in 1978 numerically determined the entropy of the random bond chain for $T \neq 0$, and at low enough temperature could see the first few minima. Then Doman and Williams[19] calculated the shape of the continuous finite temperature curves for energy, entropy, specific heat and magnetisation in the vicinity of the critical fields $B_c^{(r)}$. These curves for $|B - 2J/r| \ll k_B T = \beta^{-1}$ depend only on $\beta(B - 2J/r)$.

As the shape of these 'washed-out' continuous curves for $T \neq 0$ is qualitatively similar to the ANNNI-case, here we only refer to the corresponding results[19].

3. Two-Dimensional Frustrated Ising Systems

In the last chapter we have discussed several frustrated Ising chains
with different behavior of the entropy and the pair correlation function. These one-dimensional systems of course could not yet show a
phase transition at finite temperature, because for Ising systems
with short range interactions the lower critical dimension is $d_U = 1$,
below which thermal fluctuations destroy any long range order (LRO) .
We add, that for systems with multicomponent spins (n > 1) because
of the occuring continuous rotation symmetry the lower critical dimension is $d_U = 2$ (Mermin and Wagner 1966)[26].

In this chapter we now consider a number of two-dimensional frustrated
Ising systems which in contrast to the universal behavior in the nonfrustrated case exhibit different types of phase transitions; for
special ratios of the competing interactions some may show no or several successive transitions.

3.1 Transformations of Ising Systems

As the properties of frustrated systems depend on the detailled lattice structure, in Table 3.1 the hexagonal and in Table 3.2 the
square lattices are shown, to which most of this chapter is devoted.
This choice of lattices is not arbitrary, but suggested by the connection between the lattices in both tables given by the combination
of the following three transformations. These transformations

        (a)   duality transformation,

        (b)   decoration-iteration transformation,

        (c)   star-triangle transformation

have been introduced by Kramers and Wannier[27] (a) , Syozi[28] (b) ,
and Onsager[29] (c) . A good review gives Syozi[30] (1972). These
transformations will be briefly discussed, before we consider the
individual frustrated Ising systems in detail.

Table 3.1

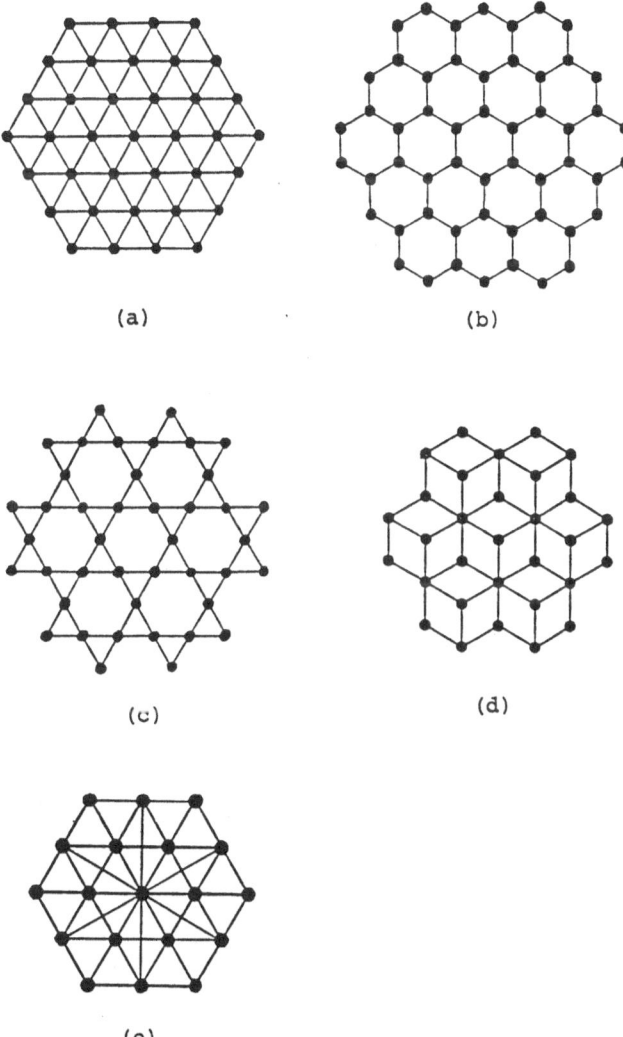

(a)

(b)

(c)

(d)

(e)

(a)   Triangular lattice
(b)   Hexagon lattice
(c)   Kagomé lattice
(d)   Diced lattice
(e)   Triangular lattice with nnn-interactions (only partially drawn)

Table 3.2

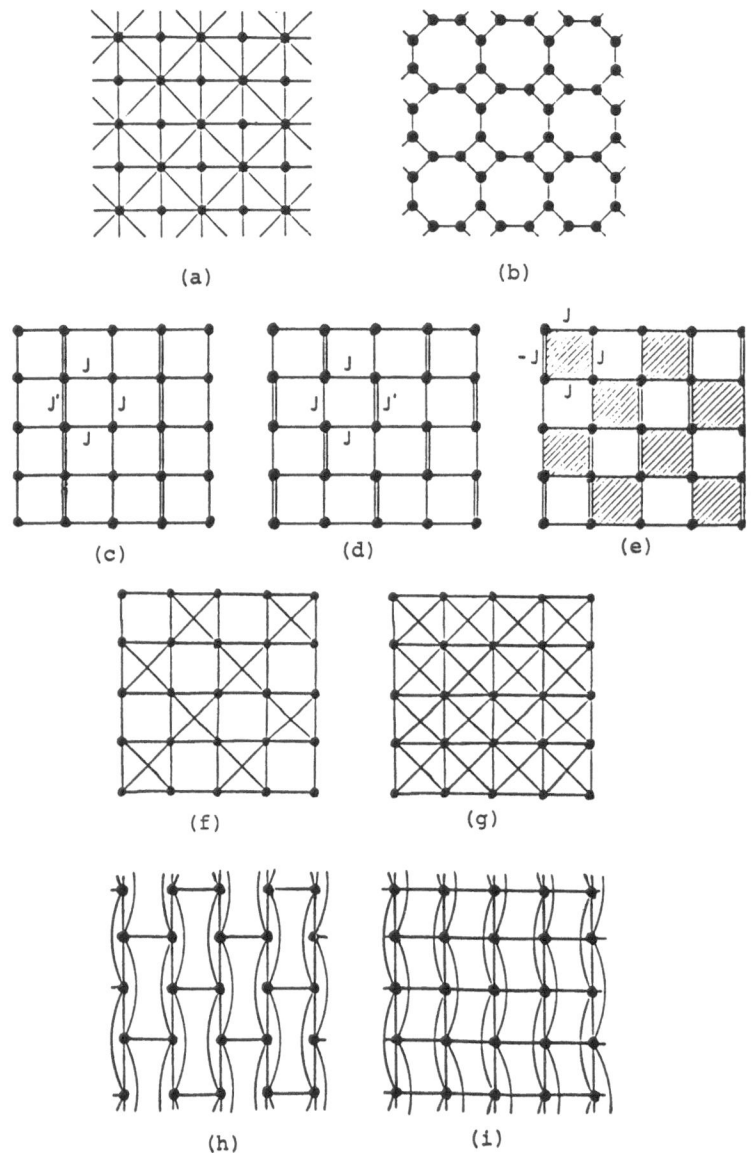

(a)

(b)

(c)

(d)

(e)

(f)

(g)

(h)

(i)

(a)   Union-Jack lattice   (f) and (g)   Square lattices with crossing
(b)   4-8   lattice   interactions
(c)   PUD   model   (h)   Brick model
(d)   ZZD   model   (i)   ANNNI-model
(e)   Chessboard model

3.1.1 Duality transformation

The duality transformation gives a connection between the partition
functions of pairs of Ising systems dual to each other, if these systems have no crossing interactions. The transformation consists of a
geometrical and an algebraical part.

α) Geometrical part: dual lattices

For a given lattice its dual is constructed by marking (by crosses) the centres of the elementary polygons and then connecting
them (by dashed lines); see Fig. 3.1a,b. From these examples it
is evident that always pairs of dual lattices occur. When both
are identical as the square lattices in Fig. 3.1a, the lattice
is called self-dual.

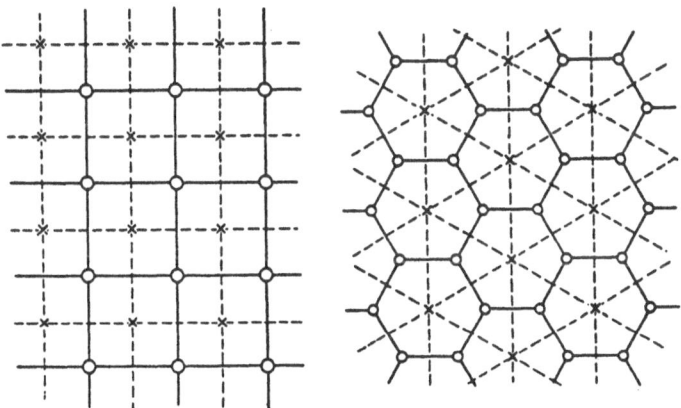

Fig. 3.1a,b Construction of the dual lattice (for d = 2). (from ref. 30).

The first four lattices in Table 3.1 are two pairs of dual lattices: triangular and hexagon lattice form one pair, Kagomé and
diced lattice form the other one. Of the second pair, the Kagomé
lattice is of special interest experimentally[31] and theoretically (as two-dimensional analogue of the three-dimensional pyrochlore lattice, Sec. 4.3).

In the triangular lattice with nnn-interaction these are crossing
each other and the nn-interactions and, therefore, there is no
dual lattice to it.

The first two lattices in Table 3.2, the Union Jack and the 4-8 lattice also form a dual pair.

The three square lattices with differing distribution of F and AF nn-interactions can also be treated exactly and show essentially different behavior.

The Ising square lattice with crossing nnn-interactions in every second square can be mapped on the 16-vertex model, solvable exactly for special cases. This is no longer possible, when as in III.2g nnn-interactions exist in every square.

Finally the brick and the uniaxial ANNNI-models are shown. The first one can be solved exactly and is interesting as comparison to the ANNNI-model for which an extensive literature exists.

β)   Algebraic part: dual partition functions

The partition function of a 2d Ising system with noncrossing nn-interactions $K(b)$ $\left(K(b) = J(b)/k_BT \right.$, b is an index for the interaction$\left. \right)$

$$Z(k) \quad = \quad \sum_{\{\sigma=\pm 1\}} \; \exp \left(\sum_b K(b) \; \sigma_i \; \sigma_j \right) \qquad (3.1)$$

for nonfrustrated systems (for instance when all $J(b) > 0$) is proportional to the partition function of an Ising system on the dual lattice with new nn-interactions $K^*(b) = J^*(b)/k_BT$,

$$e^{-2K^*(b)} \quad = \quad \tanh K(b) \qquad . \qquad (3.2a)$$

Equation (3.2a) can be rewritten in several ways:

$$e^{-2K(b)} \quad = \quad \tanh K^*(b) \qquad , \qquad (3.2b)$$

$$\sinh 2K(b) \sinh 2K^*(b) \quad = \quad 1 \qquad . \qquad (3.2c)$$

Interestingly Eq. (3.2) connects the interactions of the dual systems locally, i.e. this transformation is applicable for arbitrary inhomogeneous distributions of ferromagnetic interactions, not only in the homogeneous case.

The partition functions of homogeneous dual systems for large lattice size are linked by[30]

$$\frac{Z(k)}{2^{N/2}(\cosh 2K)^{S/2}} = \frac{Z^*(k^*)}{2^{N^*/2}(\cosh 2K^*)^{S/2}} \quad , \qquad (3.3)$$

where N and N* are the numbers of lattice sites in the original and its dual system, and S = N + N* .

If a lattice is self-dual, as the square lattice, one has N = N* . If also the dual interactions are of the same type as the original ones (for instance nn-pair interactions in 2d systems), in addition, $Z^*(k) = Z(k)$ holds. Then Eq. (3.3) simplifies to

$$\frac{Z(k)}{(\cosh 2K)^N} = \frac{Z(k^*)}{(\cos 2K^*)^N} \quad . \qquad (3.4)$$

In such self-dual Ising systems singularities in $Z(k)$ can occur either in pairs connected by Eq. (3.2) or as a special case for K = K* .

In this case Eq. (3.2) yields

$$\sinh 2K_c = 1,$$

or transformed

$$K_c = (1/2) \ln (1+\sqrt{2}) \simeq 0.4407 \quad . \qquad (3.5)$$

This is the exact inverse transition temperature of the ferromagnetic nn Ising system on the square lattice. The result can easily be generalized to the anisotropic case with different horizontal and vertical interactions J_1 and J_2[80] :

$$\sinh 2K_{1c} \sinh 2K_{2c} = 1 \quad . \qquad (3.6)$$

When antiferromagnetic interactions are present, one has to distinguish between lattices without or with frustration. In the first case the system (without external field) can be transformed by local gauge transformations to a purely ferromagnetic system, leaving the partition function invariant[3].

In frustrated systems this is not possible; the dual interactions become complex (see Eq. (3.2a,b)). As a consequence the properties of frustrated Ising systems with real interactions on dual lattices are not linked as closely as in the nonfrustrated case. This is a further hint at the different behavior of frustrated Ising systems of the same spacial dimension.

The duality transformation has been generalized by F.J. Wegner[32] to higher-dimensional Ising systems with multispin interactions, where e.g. products of 2^n spins occur in the Hamiltonian.

Refering to the self-dual Ising system with four-spin interactions on the fcc-lattice discussed in Chapter 4, we point out here already that Wegner's paper contains also the case of high ground-state degeneracy caused not by frustration, but by local gauge symmetry.

To explain this local symmetry, in Fig. 3.2a,b the unit cells of two systems $(M_{22}$ and M_{32} in the nomenclature of ref. 32) with four-spin interactions are shown.

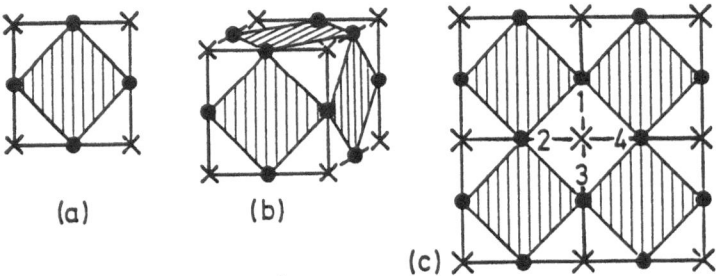

(a) (b) (c)

Fig. 3.2 (a), (b): Two-, three-dimensional lattice gauge model of Wegner[32].
The spins (•) are in the middle of the edges of the square-respecitve cubic lattice, whose sites are marked by crosses (x).
The four-spin interactions are drawn as hatched squares.
(c) : Local symmetry: Flipping spins 1 to 4 for d = 2 leaves the Hamiltonian invariant, as two spins change sign in each four-spin interaction.

From Fig. 3.2c the invariance of the Hamiltonian under the flipping of convenient clusters of spins is evident. Thus one obtains a twofold degeneracy of each state, not only of the groundstate, per marked lattice site.

As there are two respective three spin sites per marked lattice site in the two- respective three-dimensional case, the groundstate degeneracy for the models M_{22} and M_{32} with N spins (without boundary effects) is $2^{N/2}$ and $2^{N/3}$ and the corresponding groundstate entropy per spin is

$$S_o^{(22)} = \frac{1}{2} \ln 2 \qquad\qquad (d = 2) \qquad\qquad\qquad (3.7)$$

and

$$S_o^{(32)} = \frac{1}{3} \ln 2 \qquad\qquad (d = 3) \qquad . \qquad\qquad (3.8)$$

For comparison the entropy for $T \to \infty$ for arbitrary Ising systems is

$$S_\infty = \ln 2 \qquad .$$

The model M_{32} is dual to the usual Ising model with nn-interactions on the simple cubic lattice[32]. Therefore, it has like its dual a phase transition of second order at a finite T_c and it is a remarkable first example for a system which in spite of a finite groundstate entropy shows a transition with $T_c \neq 0$.

The Ising systems with multispin interactions introduced by Wegner are of such physical interest, as they are the simplest lattice gauge models; in quantum chromodynamics similar models with more general spin degrees of freedom play an important role.

The duality transformation can also be generalized to spins with several components; e.g. for XY-models $(n = 2)$ [33], but here we don't discuss this further.

3.1.2 Decimation Transformations

Already in the last chapter we have talked of simple and multi-decorated
chains (see Fig. 2.1), where for computing the partition function one
can sum exactly over part of the spins. One obtains a temperature dependent effective interaction between the remaining spins. This method
is not restricted to one-dimensional systems, but can be applied in
all cases, where single spins or groups of spins interact with only
two spins of the remaining system (ref. 34). When they interact with
three or four spins of the remaining system, on obtains also new three-
and four-spin interactions.

3.1.2a Decoration-Iteration Transformation

We first consider the simplest case shown in Fig. 3.3a, where a central spin interacts with two neighboring spins. For different original
interactions K_1 and K_2 one obtains the effective interaction[34]

$$K' = \frac{1}{2} \ln \left(\frac{\cosh (K_1 + K_2)}{\cosh (K_1 - K_2)} \right) \quad , \tag{3.9}$$

or more symmetrical

$$\tanh K' = \tanh K_1 \tanh K_2 \quad . \tag{3.10}$$

For $K_1 = K_2 = K$ this yields again $K' = 1/2 \ln \cosh 2K$, the in-
direct contribution to the nnn-interaction of the simple decorated
chain, Eq. (2.25). This transformation can always be inverted. However, the inversion is not unambiguous because in a chain without
direct nnn-interactions frustration cannot occur.

The decoration-iteration transformation can be generalized to several
intermediate spins, also to a magnetic field (ref. 34), as long as it
does not act on the intermediate spins.

We discuss applications of this transformation in the now following
section on the star-triangle transformation.

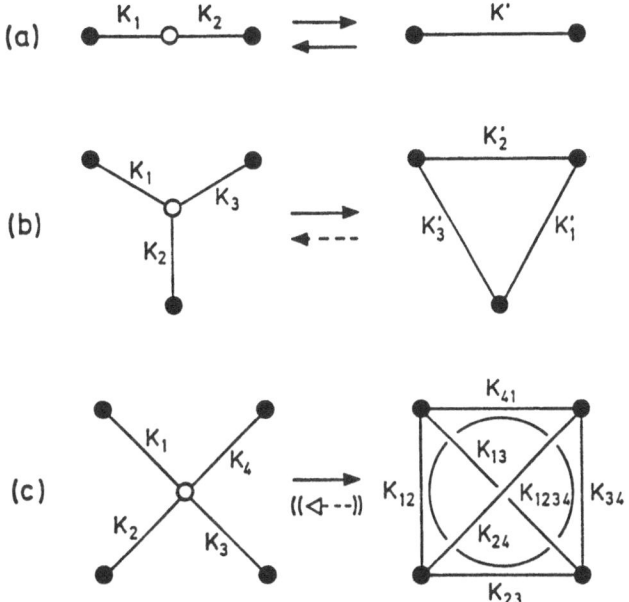

Fig. 3.3   Decimation transformations:
   (a)   Decoration-iteration transformation,
   (b)   star-triangle transformation,
   (c)   star-square transformation.

3.1.2b   Star-Triangle_Transformation

This transformation is completely analog to the decoration-iteration transformation, only the intermediate spin or group of spins is now interacting with three other spins. Figure 3.3b again shows the simplest case. Without a magnetic field the transformation from the star interactions (no primes) to the triangle interactions (primes) is[30,34]:

$$\psi_0 = \cosh\ (K_1 + K_2 + K_3)$$

$$\psi_1 = \cosh\ (-K_1 + K_2 + K_3)$$

$$\psi_2 = \cosh\ (K_1 - K_2 + K_3) \tag{3.11}$$

$$\psi_3 = \cosh\ (K_1 + K_2 - K_3)$$

$$e^{4K_1'} = \frac{\psi_0\ \psi_1}{\psi_2\ \psi_3}\ , \qquad e^{4K_2'} = \frac{\psi_0\ \psi_2}{\psi_3\ \psi_1}\ , \qquad e^{4K_3'} = \frac{\psi_0\ \psi_3}{\psi_1\ \psi_2}\ .$$

When all star interactions are identical $(K_i = K)$, Eq. (3.11) becomes

$$e^{4K'} = \frac{\cosh\ 3K}{\cosh\ K}\ . \tag{3.12}$$

Independent of the sign of K the new triangle interactions are ferromagnetic $(K' > 0)$. The inverse transformation is possible[30], from (3.12) one realizes, however, that for a frustrated triangle $(K' < 0)$ the corresponding star interaction K becomes imaginary. Similar to the duality transformation also the star-triangle transformation maps Ising Hamiltonians with real interactions on the real Ising Hamiltonians only, when no frustration is present.

In addition in Fig. 3.3c the star-square transformation is shown, where the central spin interacts with four neighboring spins. This transformation generates seven new interactions K_i' from the four original ones K_i . Two of the new ones are nnn , one is a four-spin interaction. As the seven new interactions depend on the four original ones, they are not independent. Therefore, this transformation can be inverted only for special cases and is of minor importance.

3.1.3 Connection Between the Different Lattices

The decimation transformations in addition to duality yield further connections between the lattices depicted in Tables 3.1 and 3.2. Application of the inverse star-triangle transformation to half of the triangles of the triangular lattice (Fig. 3.4) leads to the hexagon lattice, to all triangles to the diced lattice.

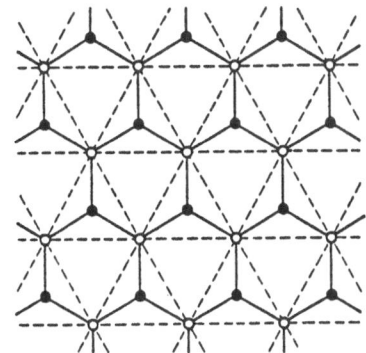

Fig. 3.4 Transformation between triangular and hexagon lattice.

After decoration of all bonds of the hexagon lattice with an intermediate spin the star-triangle transformation leads to the Kagomé lattice. The connections derived from duality and decimation transformations between the various hexagon lattices of Table 3.1 are summed up in Fig. 35 [30].

As to the square lattices of Table 3.2, using the star-square transformation one can show the triangular lattice, Villain's odd model and the Union Jack lattice to be special cases of Baxter's 8-vertex model[35,36] which will be further discussed in Section 3.3.4.

In the following sections we discuss the two-dimensional Ising systems mentioned in more detail, with special regard of the frustration effects.

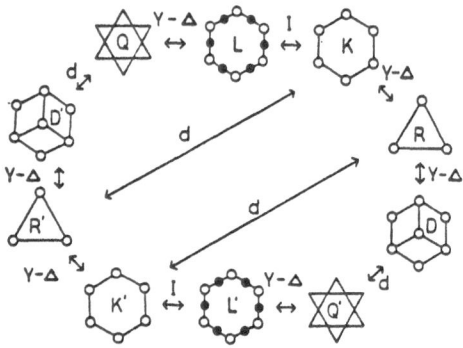

Fig. 3.5 Connection between the hexagon lattices of Table 3.1, derived from
duality (d) , star‑triangle (Y-Δ) and decoration-iteration (I)
transformations (ref. 30).

3.2 Triangular Lattice

The antiferromagnetic Ising system on the triangular lattice, the
first frustrated system, has been investigated by Wannier[37] and Houtappel[38] using transfer-matrix methods already a few years after Onsager's exact solution of the ferromagnetic square lattice. They determined the partition function and several related thermodynamic quantities, and also found the transition temperature T_c (≥ 0) . Wannier
only considered the case of isotropic nn-interactions, whereas Houtappel treated the anisotropic case, where the three differently oriented nn-interactions have different values (Fig. 3.6).

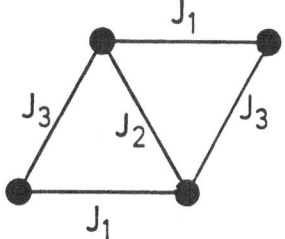

<u>Fig. 3.6</u> Unit cell of the anisotropic triangular lattice.

3.2.1 Estimation of the Groundstate Degeneracy

In the parameter space J_1, J_2, J_3 of the three interactions it is sufficient to consider the groundstate (GS) properties on a unit sphere,
as they are independent of a common scaling factor (> 0) . Also the
simultanuous change of sign of two interactions leaves the thermodynamic quantities invariant (in the absence of a magnetic field), because
only every second row of spins oriented in the direction of the third,
fixed interaction is flipped; always four points $(J_1 J_2 J_3)$, $(J_1,-J_2,-J_3)$,
$(-J_1,-J_2,J_3)$ are equivalent. Therefore, it is sufficient to consider
only the spherical triangle in Fig. 3.7 with the corners AF_1, AF_2 and
AF_3 , containing one fourth of the surface. Its contours are defined
by $J_1 + J_2 = 0$, $J_2 + J_3 = 0$ and $J_3 + J_1 = 0$. Inside this triangle
the GS is ferromagnetic (F) . The point F marks the isotropic
ferromagnet. Along the dashed lines one of the three interactions vanishes corresponding to anisotropic square lattices. The points Q_i
mark the isotropic square lattices.

Along the edges (full lines), e.g. AF_1 - K_3 - AF_2 one interaction
(here J_3) dominates over the other two of equal absolute value (here
$J_2 = -J_1$) . The GS's consist of ideally ordered 1d-chains (here in
J_3-direction), without order between different chains. At the points
K_i these chains are exactly decoupled also for $T \neq 0$. Though the

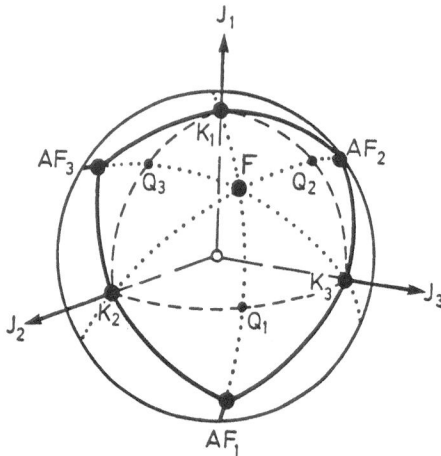

Fig. 3.7 Parameter space of the anisotropic triangular lattice.

Special points are:

F : (1/√3) (1,1,1) isotropic ferromagnet (F) on triangular lattice,

Q_i : e.g. (1/√2) (0,1,1) isotropic F on square lattice,

K_i : e.g. (1,0,0) 1d F chain, $T_C = 0$,

AF_i : e.g. (1/√3) (1,1,$\overline{1}$) frustration points, $T_C = 0$, corresponding to isotropic AF on triangular lattice.

GS is degenerate, in the thermodynamic limit N → ∞ the GS entropy per site vanishes proportional to $N^{-1/2}$.

Here we are mainly interested in the (equivalent) corners AF_i of the spherical triangle, which are the frustration points of the system. All three interactions have the same absolute value, and each elementary triangle is fully frustrated. Already for a single triangle this leads to a high GS entropy (six of the eight states are GS). In the infinite system the GS degeneracy is high enough to yield a finite GS entropy S_o per site, as will be discussed in more detail below.

Before considering the exact determination of S_o by Wannier, we first turn to three ways to estimate S_o which are applicable also for other systems, for which no exact results are available (especially for d = 3) .

3.2.1a Simple Lower Bound

In his paper Wannier discusses different classes of GS with different statistical weight. As examples in Fig. 3.8a,b,c GS with entropies per lattice site $S_o \propto N^{-1}$, $N^{-1/2}$, const. are shown, where N is the number of sites. The configuration c has the highest weight, and one notes that at least every third spin (all spins on one of the three sublattices) is completely free. This immediately yields a lower bound for S_o:

$$S_o > \frac{1}{3} \ln 2 \approx 0.2310 \quad ,$$

which can be raised to

$$S_o > \frac{5}{12} \ln 2 \approx 0.2888 \tag{3.13}$$

taking into account nn-corrections[37].

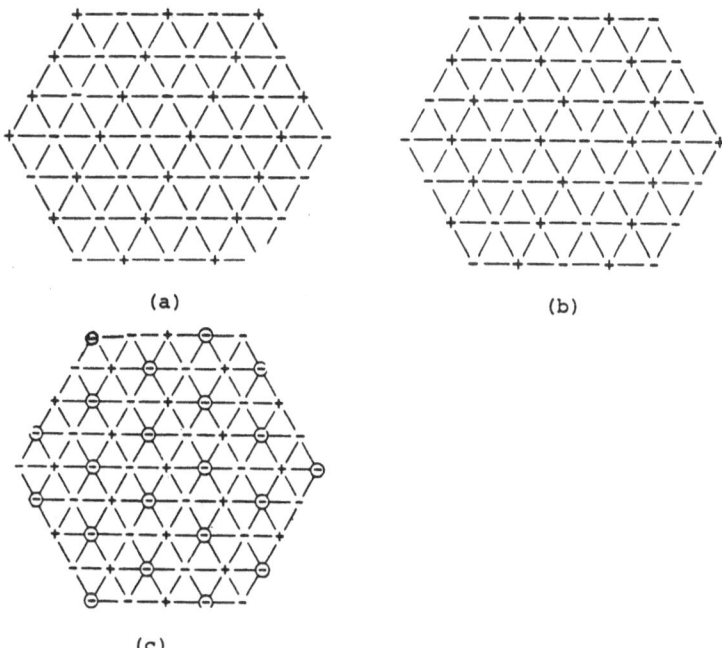

(a)

(b)

(c)

Fig. 3.8 Three types of groundstates in the AF   triangular lattice:
(a) AF chains periodically stacked,
(b) AF chains random stacked,
(c) $\sqrt{3} \times \sqrt{3}$ structure, two of three sublattices are F ordered, the spins on the third (o) are completely free.

3.2.1b Pauling_Method

A second approximation which usually is neither an upper nor a lower
bound, is the method of Pauling[39]. Here the number of GS N_g is
estimated by multiplying reduction factors to the total number of
states, 2^N. These factors are the probabilities to find the elementary
triangles of the lattice in GS configurations. As in an AF triangle
six of eight states are GS , $w_o(3) = 6/8$. There are two triangles per
site, hence

$$S_o = \lim_{N \to \infty} \frac{1}{N} \ln N_g \simeq \lim_{N \to \infty} \frac{1}{N} \ln \left(2^N \, w_o(3)^{2N} \right) \quad . \qquad (3.14)$$

In this approximation N drops out, and one obtains

$$S_o(3) = \ln 2 + 2 \ln w_o(3) \simeq 0.118 \quad . \qquad (3.15)$$

This is by far too low, in this simple form the method is unsatisfac-
tory for lattices with high coordination number. However, this method
yields very good results for the Kagomé lattice and for ice models.

3.2.1c Systematic_Cluster_Approximation

One obtains a much better approximation for S_o by covering the lattice
not only with elementary triangles, but with clusters of different
size M , calculating the GS probability for these clusters $w_o(M)$
exactly and finally extrapolating the corresponding $S_o(M)$ to
M → ∞ . This is a simple form of 'finite size scaling'[40-42]. As the
deviations

$$\Delta S_o(M) = S_o(M) - S_o(M \to \infty) \qquad (3.16)$$

are an effect of the cluster boundary, one can expect $\Delta S_o(M)$ to be
proportional to the relative number of boundary atoms

$$\Delta S_o(M) \propto \frac{\sqrt{M}}{M} = M^{-1/2} \quad . \qquad (3.17)$$

Therefore, for extrapolation to M → ∞ it will be useful to plot
$S_o(M)$ versus $M^{-1/2}$, as one should find then an approximately linear
behavior.

We have considered two ways to cover the lattice by clusters which are shown in Fig. 3.9 for (M = 6) triangular clusters and which yield different approximations for S_o .

For case (a) one obtains

$$S_o^{(a)} (M) = \ln 2 + \frac{1}{M} \ln w(M) = \frac{1}{M} \ln N_g(M) \qquad , \qquad (3.18)$$

for case (b)

$$S_o^{(b)} (M) = \ln 2 + \frac{2}{M_p} \ln w(M) \qquad , \qquad (3.19)$$

where M_p is the number of elementary triangles in the M-cluster.

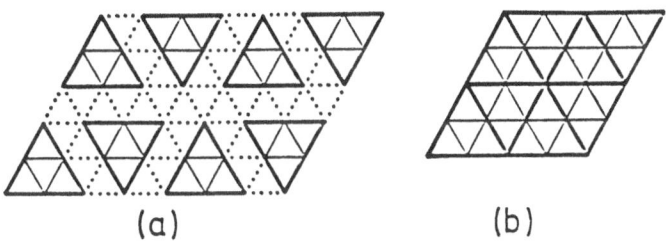

(a)   (b)

Fig. 3.9   Two ways to cover a lattice with clusters.

    (a)   Every lattice site belongs to one and only one cluster corresponding to Eq. (3.18);

    (b)   no space between clusters corresponding to Eq. (3.19).

As in the limit $M \to \infty$ the ratio $2M/M_p \to 1$, in this limit the two expressions for S_o coincide. We note that the simple Pauling method, Eq. (3.15), corresponds to $S_o^{(b)} (M = 3)$.

$S_o^{(a)} (M)$ is an upper bound, as one overestimates the number of ground-states when only the triangles inside the different clusters are re-quired to be in triangle GS .

$S_o^{(b)} (M)$ is always lower than $S_o^{(a)} (M)$, as the clusters are more densely packed, but it need not be a lower bound.

We have determined the GS degeneracy $N_g(M)$ and thus also $w_o(M)$ for triangular $(M = 3, 6, 10, 15, 21)$, rhombic $(M = 4, 9, 16, 25)$ and hexagonal $(M = 7, 21)$ clusters, the first two clusters of each shape are shown in Fig. 3.10.

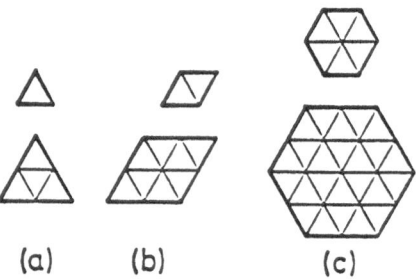

(a)   (b)   (c)

<u>Fig. 3.10</u> The first two clusters of each shape used for extrapolation.

We note that $N_g(M)$ is determined here by requiring two spins up and one down or the reverse for each elementary triangle of the cluster. This is the true GS of a finite cluster only, when the interactions along the edge of the cluster are reduced by a factor of two. However, this way the dependence of N_g on the boundary is reduced significantly and better values for the extrapolated $S_o(\infty)$ are obtained. Also only using this way to calculate $N_g(M)$, $S_o^{(a)}(M)$ is necessarily an upper bound.

In Fig. 3.11 the entropies $S_o^{(a)}$ and $S_o^{(b)}$ are plotted versus $M^{-1/2}$. $S_o^{(a)}(M)$ is very close to linear in $M^{-1/2}$ within the three series except for the very smallest clusters $(M \leq 7)$. The extrapolated value $S_o^{(a)}(\infty)$ agrees with the exact result (full circle) to within 3 percent. $S_o^{(b)}(M)$ is more sensitive to cluster shape than $S_o^{(a)}(M)$, but extrapolation within each series also agrees with the exact result within similar error bars.

We add that $\Delta S_o^{(i)}(M)$ (especially for $i = b$) is largest for triangular and smallest for hexagonal clusters. This is also a cluster edge effect, the triangular clusters possessing the largest, and the hexagonal ones the smallest relative number of edge sites.

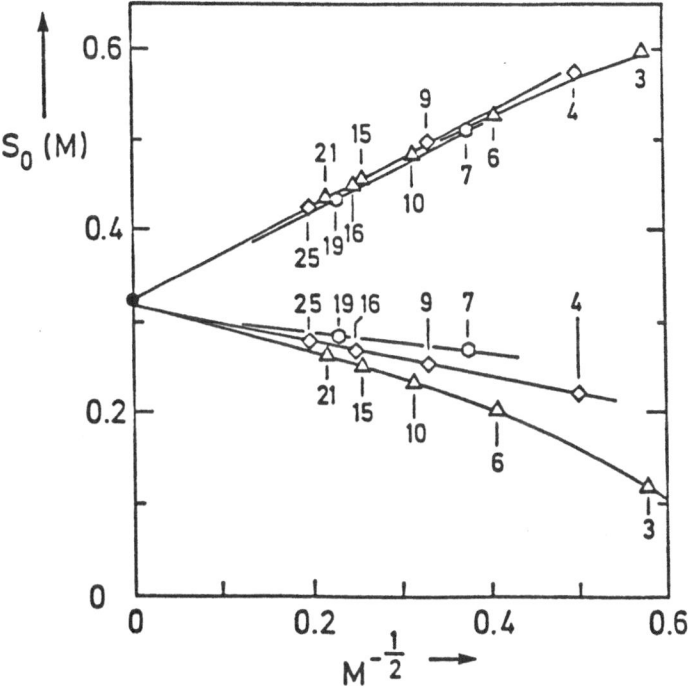

Fig. 3.11 $S_0(M)$ for M > 7 is to good approximation linear in $M^{-1/2}$. Extrapolation to M → ∞ agrees to within 3 percent with the exact value (•) .

The shape dependent part of edge sites for M → ∞ is

$$C_E = \alpha_E \cdot M^{-1/2} \qquad , \qquad (3.20)$$

with $\alpha_E = 3\sqrt{2}$, 4 , $6/\sqrt{3} \simeq 4.24$, 40, 3.46 for triangular rhombic, and hexagonal clusters.

This method to calculate the GS entropy yields quite precise results for small cluster sizes and can be augmented systematically using larger clusters.

3.2.2 Partition Function and Exact Groundstate Entropy of the Isotropic System

As already mentioned, Wannier[37] and Houtappel[38] have calculated exactly the partition function and thus also internal energy and entropy of the frustrated triangular lattice. Houtappel obtained for the partition function per site λ ($Z = \lambda^N$, $F = - kT \ln \lambda$) :

$$\ln \left(\frac{\lambda}{2}\right) = \frac{1}{8\pi^2} \int_0^{2\pi} \int_0^{2\pi} \ln \left(C_1 C_2 C_3 + S_1 S_2 S_3 - S_1 \cos\omega_1 - S_2 \cos\omega_2 - S_3 \cos(\omega_1+\omega_2) \right) d\omega_1 \, d\omega_2 \quad,$$

(3.21)

with

$$C_i = \cosh 2K_i \quad, \qquad S_i = \sinh 2K_i \quad.$$

From this result the internal energy $U(T)$, the entropy $S(T)$, the specific heat $c(T)$ and T_c can be determined.

In the anisotropic system for the parameter range corresponding to the spherical triangle of Fig. 3.7 one obtains for T_c (with $S_i^c = \sinh (2J_i/kT_c)$):

$$S_1^c S_2^c + S_2^c S_3^c + S_3^c S_1^c = 1 \quad.$$

(3.22)

Equation (3.22) contains the well known case of the nonfrustrated anisotropic square lattice; only one of the three interactions has to be set to zero:

$$\sinh 2K_x^c \sinh 2K_y^c = 1 \quad.$$

(3.23)

When the spherical triangle of Fig. 3.7 is mapped on the plane $J_1 + J_2 + J_3 = $ const. , Eq. (3.22) yields the lines of constant $T_c/(J_1+J_2+J_3)$ shown in Fig. 3.12a. One notes $T_c \neq 0$ whenever the GS has two-dimensional long range order (LRO) . Along the lines $AF_i - K_{i+2} - AF_{i+1}$, where the GS has only 1d LRO , T_c vanishes although there is not yet a finite GS entropy per lattice site.

The vanishing of T_c along the lines $AF - K - AF$ is plausible, as crossing these lines the GS changes, causing a higher degeneracy right on the lines. However, later we shall see (for the square lat-

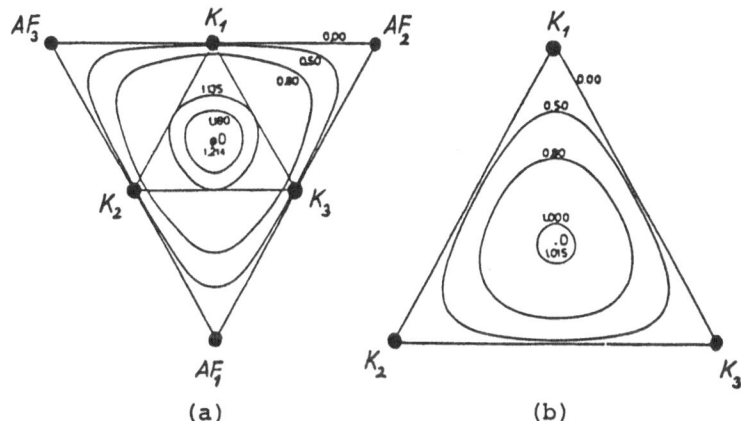

Fig. 3.12 (a) : Triangular lattice. Lines of constant transition temperature
$T_C / (J_1+J_2+J_3)$ (after ref. 38).

  (b) : Hexagon lattice (not frustrated in this parameter space). Lines
of constant $T_C/(J_1+J_2+J_3)$. Here T_C vanishes only when at
least one interaction J_i vanishes, as the lattice then falls
apart into seperate 1d chains (after ref. 38).

tice) that even for finite GS entropy per site T_C may be finite;
the properties $S(T=0) \neq 0$ and $T_C = 0$ need not occur simultaneously.

For comparison in Fig. 3.13 the internal energy $U(T)$ for the two
isotropic cases is shown, where all $|J_i|$ are equal corresponding to
the points F_i respectively AF_i of Fig. 3.12.

Whereas in the nonfrustrated case there is no essential difference to
the ferromagnetic square lattice and a normal Ising transition occurs
at a finite T_C , the antiferromagnet exhibits no singularity for
$T \neq 0$. Because of the frustration the GS energy $U(0)$ is raised
to $-0.5 |J|$ from $-1.5 |J|$. From Eq. (3.21) one obtains the exact
value for the GS entropy S_0 at the frustration points[37]:

$$\frac{S_0}{K_B} = \frac{3}{\pi} \int_0^{\pi/6} \ln (2\cos\omega) \, d\omega \simeq 0.323066 \qquad . \qquad (3.24)$$

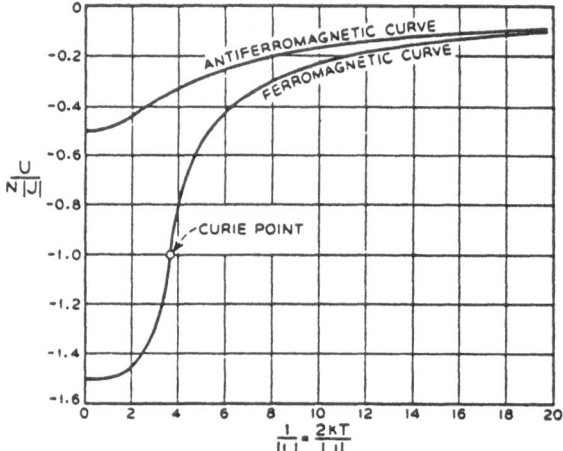

<u>Fig. 3.13</u> Internal energy U(T) of the isotropic ferromagnet and antiferromagnet
on the triangular lattice (ref. 37).

This value was used already in Fig. 3.11 for comparison with the finite
cluster results.

3.2.3 Specific Heat Near the Frustration Points $(J_1 = J_2)$

Vaks and Geilikman[43] have studied the behavior of the frustrated triangular Ising system near the frustration points for the case $J_1 = J_2 = J$,
$J_3 = J + \delta$, $J < 0$, $|\delta| \ll |J|$. Then two cases have to be distinguished:

  1.) $\delta > 0$: the GS shows 2d LRO ; $T_c \neq 0$;

  2.) $\delta < 0$: the GS shows 1d LRO only; $T_c = 0$.

In Fig. 3.14 the specific heat c(T) is shown as full line for $\delta > 0$
and as dashed line for $\delta < 0$.

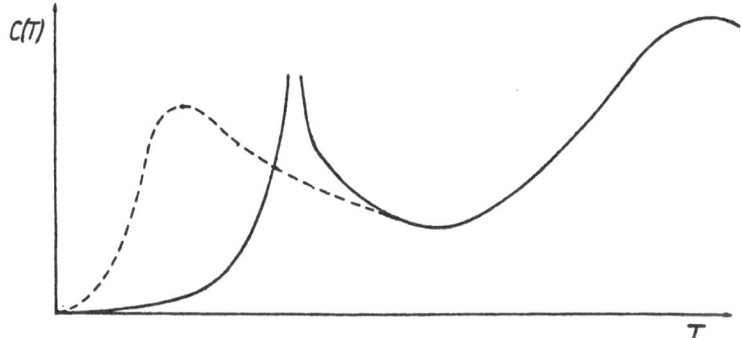

Fig. 3.14 Specific heat of the anisotropic AF triangular lattice $(J_1 = J_2 = J_3 - \delta ,$
$|\delta| << |J_i|)$. For $\delta > 0$ (full curve) there is a logarithmic divergence,
not for $\delta < 0$ (dashed curve). (after ref. 43).

As expected the specific heat is independent of δ in the range
$T >> |\delta|$:

$$c \propto |K|^3 e^{-4|K|} . \qquad\qquad (3.25)$$

There is a broad maximum at $T_m \simeq |J|$. For positive $\delta << |J|$ an
Ising transition occurs at the much lower temperature $T_c \simeq 2\delta/\ln 2$,
with typical logarithmic divergence in the critical range near T_c .
Further above respectively below T_c one finds $t^{-1/2}$ respectively
$t^{-3/2}$ power law behavior, which is quite uncommon.

For $\delta < 0$ there only occurs an analytic maximum at $T \simeq |\delta|$, because
of the absence of 2d LRO at any temperature. The low T maximum
corresponds to SRO . For $T > |\delta|$ the interactions of strength J
and $J + \delta$ are broken with almost equal probability, for $T < |\delta|$
almost only the weaker ones. But this even for T = 0 only leads to
1d LRO (see the next Sec. 3.2.4).

For comparison Vaks and Geilikman discuss the specific heat of the extremely anisotropic ferromagnet on the square lattice, Fig. 3.15. The
interactions J_1 in vertical direction are supposed to be much weaker
than the horizontal ones J_2 , $J_1/J_2 = \gamma << 1$. For $\gamma = 0$ the system
decomposes into independent horizontal chains; the corresponding specific heat

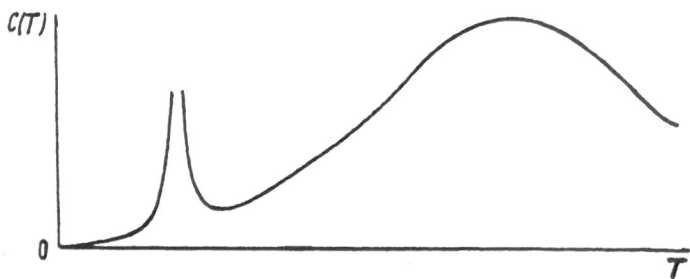

Fig. 3.15 Specific heat of the extremely anisotropic ferromagnet on the square lattice (ref. 43).

$$c(T) \;=\; K_2^2 \; \cosh^{-2} K_2 \qquad\qquad (3.26)$$

has a broad maximum around $T_m \simeq J_2$. For a small additional J_1 ($\gamma \ll 1$) the specific heat in the range of T_m is unchanged compared to $J_1 = 0$. Interchain order begins to build up only at much lower temperature for $\gamma \ll 1$, and the Ising transition is at $T_c \simeq 2\, J_2/\ln(1/\gamma)$. Above T_c there is a crossover from 2d to 1d behavior. Near T_c one obtains for $c(T)$ [43]:

$$c(T) \;\simeq\; \frac{9\gamma}{4\pi} \, (\ln\gamma)^2 \, \ln \, (1/|t|) \qquad ; \qquad\qquad (3.27)$$

the logarithmic divergence is much weaker compared to the isotropic 2d ferromagnet, and the temperature dependence is comparable to the case $\delta > 0$ of the triangular lattice (Fig. 3.14).

3.2.4 Pair Correlation Function, Disorder Lines $(J_1 = J_2)$

Leaving the thermodynamic functions we now consider the pair correlation function $G(r)$ containing detailled information on the spin order. Stephenson[44] has calculated the pair correlation function for the
general anisotropic case.

Here we discuss as in Section 3.2.3 only the case $J_1 = J_2 = J < 0$,
$J_3 = J_1 + \delta$ or $J_3/J_1 = \lambda$ and like there distinguish the three cases
$\delta > 0$, $\delta = 0$, $\delta < 0$ respectively $\lambda < 1$, $\lambda = 1$, $\lambda > 1$.

For $\lambda < 1$ there is a transition at a finite T_c (see Fig. 3.14).
Above T_c along a disorder line T_D (dashed line in Fig. 3.16) $G(r)$
decays exactly exponentially; along the (1)- and (2)-directions

$$G_{1/2}(r) = (\tanh K_1)^r = (-1)^r e^{-r/\xi_{1/2}(T)} , \qquad (3.28)$$

along the (3)-direction

$$G_3(r) = (-\tanh K_3)^r = e^{-r/\xi_3(T)} . \qquad (3.29)$$

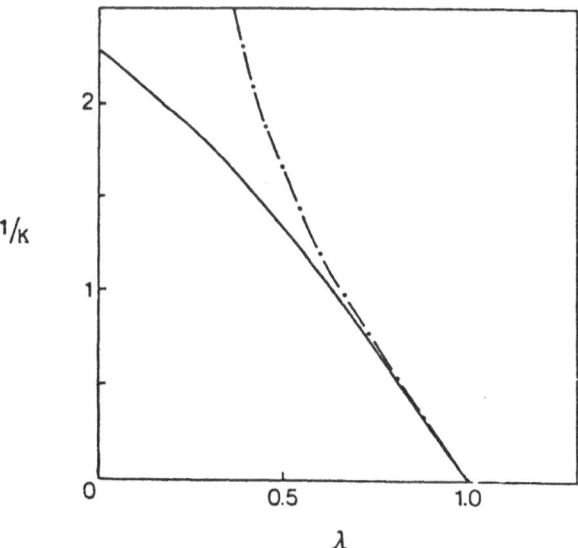

Fig. 3.16 Phase diagramm of the anisotropic AF triangular lattice. Full line: T_c ,
dashed line: disorder line T_D (after ref. 36).

with

$$\xi_{1/2}^{-1}(T) \quad = \quad \ln |\tanh K_1| \quad ; \quad \xi_3^{-1}(T) \quad = \quad 2\xi_{1/2}^{-1} \quad . \quad (3.30)$$

This pure exponential decay of $G(r)$ corresponds exactly to $G(r)$ in a 1d ferromagnetic Ising chain. Along the disorder line the triangular system is effectively one-dimensional. The disorder line ends at the isotropic frustration point $(\lambda = 1, T = 0)$.

Below T_D $G_{1/2}(r) \cdot (-1)^r$ and $G_3(r)$ are monotonously decreasing functions of r , which have a finite limit for $r \to \infty$ when $T < T_c$ ($\hat{=}$ LRO) . Contrary to this for $T > T_D$ an additional characteristic factor $r^{-1/2}$ and temperature dependent oscillations occur, with wavevectors $q_{1/2}$ respective q_3 vanishing continuously for $T \to T_D$. In the asymptotic range $r \to \infty$ one has:

$$G_{1/2}(r) \quad \propto \quad (\tanh K_1)^r \cdot r^{-1/2} \cos \left(q_{1/2} \, r + \phi_{1/2} \right) \quad ,$$
$$(3.31)$$

$$G_3(r) \quad \propto \quad (-\tanh K_3)^r \cdot r^{-1/2} \cos \left(q_3 \, r + \phi_3 \right) \quad .$$

In Fig. 3.17a,b[44] the temperature dependence of $\theta_1 = q_{1/2}$ and $\theta_3 = \pi - q_3$ is shown for several values of $\lambda = J_3/J_1$.

As the wavevectors $q_i(T)$ are continuous above T_D , like for the periodic ANNNI-chain the disorder line is of the first kind. The similarity between the 2d and this 1d system becomes especially clear when comparing Fig. 3.17 to Fig. 3.18[13], where $\theta = q$ is plotted versus T (Fig. 3.18 contains the same information as Fig. 2.2, but is more convenient for comparison).

For $\lambda = 1$ (isotropic) $T_c = T_D = 0$. For $T > 0$ Stephenson[45] essentially finds asymptotically

$$G(r) \quad \propto \quad (\tanh K)^r \cdot r^{-1/2} \cos (\theta r) \quad (3.32)$$

similar as for finite T_D (Eq. (3.31)), but for $T \to 0$ one has $\theta \to \pi/3$ (see Fig. 3.17). For $T = 0$

$$G^0(r) \quad \simeq \quad 0.63222 \cdot r^{-1/2} \cos \left(\frac{2}{3} \pi r \right) \quad , \quad (3.33)$$

corresponding to the absence of LRO for $T \geq 0$, but for $T = 0$

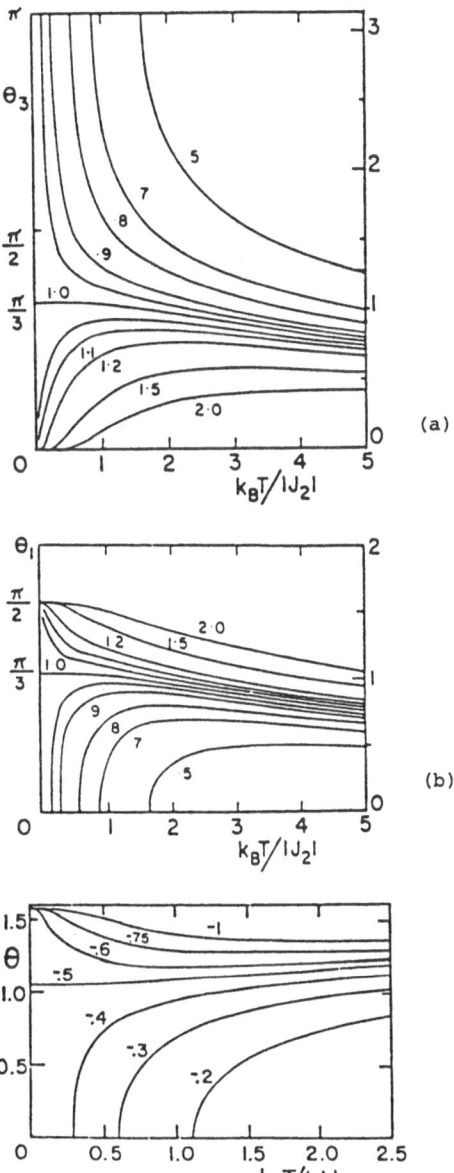

Fig. 3.17

Anisotropic triangular lattice. Continuous temperature
dependence of the wavevectors

(a) $q_3 = \pi - \Theta_3$ and

(b) $q_{1/2} = \Theta_1$ for fixed

values of J_3/J_1 (with
$J_2 = J_1$) in the disordered
phase (ref. 44).

(a)

(b)

Fig. 3.18

For comparison: 1d ANNNI-
chain. Wavevector $q = \Theta$
versus temperature for
fixed values of $J_{nnn}/|J_{nn}|$
(ref. 13).

the correlation length ξ diverges because of the power law
$G^0(r) \propto r^{-1/2}$, which is usual for $T = T_c$ at a phase transition of
second order. However, the exponent is $\eta = 1/2$, contrary to $\eta = 1/4$
found for all nonfrustrated Ising systems forming a single universality class. This $r^{-1/2}$ power law at $T = 0$ we will find also for

several, but not all other frustrated 2d systems, another hint to the complex behavior of frustrated systems.

For $\lambda > 1$ J_3 dominates over $J_1 = J_2$ and $T_c = 0$; there is no disorder line. $G(r)$ again has the form of Eq. (3.31), but $q_{1/2}(T)$ and $q_3(T)$ behave different from the case $\lambda \leq 1$.

For $T = 0$

$$G_3^0(r) = (-1)^r , \qquad (3.34)$$

that is in the GS indeed perfect 1d order occurs in (3)-direction, as was mentioned already in the last section. Along the other two directions one finds asymptotically[44]

$$G_1^0(r) \simeq 0.58835 \cdot r^{-1/2} \cos\left(\frac{\pi}{2} r\right) . \qquad (3.35)$$

After first sight one might expect the orientations of the ordered (3)-chains to be completely independent; however, this is true only for odd distance r . Chains with even distance r are correlated down to $T = 0$ and there again the factor $r^{-1/2}$ occurs. Very similar behavior we shall find again for the anisotropic frustrated square lattice. For $T > 0$ the wavevectors $q_{1/2}$ and q_3 vary continuously again (see Fig. 3.17) and approach $q_i = \pi$ for $T \to \infty$, corresponding to AF short range order.

3.2.5 Mapping to the Quantum XY-Chain and to the Kinetic nn Ising Chain

In this section we want to mention an interesting method to determine correlation functions, used by Peschel[36] for the AF triangular lattice and several other 2d Ising systems.

The mapping of 2d classical systems to 1d quantum systems is based on the fact, that thermodynamic quantities of 2d Ising systems with short range interactions can be derived from a 1d operator, the transfer matrix V . This operator can always be expressed in terms of Pauli matrices, but usually has a complicated structure because of noninterchangeable exponential factors. The matrix V becomes especially simple for two cases:

1.) One succeeds in finding a new operator $\hat{H}$,

  commuting with V ; (3.36a)

2.) it is possible to write V in the form

  $V = \exp(-H)$ (Hamiltonian limit). (3.36b)

Then it suffices to investigate $\hat{H}$ respective H which often are hermitian and which describe 1d quantum systems.

Peschel[36] has shown, how the AF triangular lattice, the Union Jack lattice and Villain's odd model can be mapped to the quantum XY-chain with special ratios of the interactions. Here we only consider the AF triangular lattice, for the other cases the procedure is analog.

First the system can be mapped to an 8-vertex model on the square lattice, satisfying the so-called 'free fermion condition' for the vertex weights $w_1, \ldots, w_8$, making possible the complete solution (Hurst and Green 1964)[46]:

$$w_1 w_2 + w_3 w_4 = w_5 w_6 + w_7 w_8 \quad . \tag{3.37}$$

When this condition is satisfied, the Hamiltonian

$$\hat{H} = - \sum_n \left\{ J_x \, \sigma_n^x - J_y \, \sigma_{n-1}^z \, \sigma_n^x \, \sigma_{n+1}^z + B \, \sigma_n^z \, \sigma_{n+1}^z \right\} \tag{3.38}$$

commutes with the transfer matrix V of the original AF triangular lattice, if[47]

$$J_x = 1 \quad ,$$

$$J_y = \tanh^2 K_3 \quad ,$$

$$B = (\sinh 2K_1 \sinh 2K_2 \cosh 2K_3 + \cosh 2K_1 \cosh 2K_2 \sinh 2K_3)/\cosh^2 K_3 \quad .$$

$$\tag{3.39}$$

Finally the Hamiltonian (3.38) can be rewritten by the dual transformation

$$\sigma_n^z \, \sigma_{n+1}^z = \tau_n^z \quad , \qquad \sigma_n^x = \tau_{n-1}^x \, \tau_n^x \quad , \tag{3.40}$$

as the Hamiltonian of the quantum XY-model in a (transverse) field,

$$
H_{XY} = -\sum_n \left\{ J_x \, \tau_n^x \, \tau_{n+1}^x + J_y \, \tau_n^y \, \tau_{n+1}^y + B \, \tau_n^z \right\} \quad , \tag{3.41}
$$

where only nn-interactions occur. Spin correlations in the horizontal direction of the original lattice can be expressed with the dual variables[36]:

$$
\begin{aligned}
G(r) &= \langle \sigma_\ell^z \, \sigma_{\ell+r}^z \rangle \\[2mm]
&= \langle \tau_\ell^z \, \tau_{\ell+1}^z \cdots \tau_{\ell+r-1}^z \rangle \\[2mm]
&= (-1)^n \, \langle \exp \left(i \, \pi \sum_{i=1}^{\ell+r-1} c_i^+ \, c_i \right) \rangle \quad ,
\end{aligned} \tag{3.42}
$$

where Fermi operators have been introduced using the Wigner-Jordan transformation.

At the transition of the original Ising system the groundstate of the corresponding XY-chain changes, that is two eigenstates cross each other as a function of the parameters. This statement contains the assumption (which can be proven) that the GS wave function of H_{XY} yields the largest eigenvalue of the transfer matrix and thus determines the thermodynamics of the original system.

In the Fermion representation H_{XY} can be diagonalized and one obtains the eigenvalue spectrum $\varepsilon(q)$ usually having a finite energy gap. But for the three cases:

(a) $J_x + J_y = -B$; $\varepsilon(0) = 0$,

(b) $J_x + J_y = +B$; $\varepsilon(\pi) = 0$, (3.43)

(c) $J_x = J_y$; $|B| \leq 2 J_x$; $\varepsilon(\Theta) = 0$,

with $\Theta = \arccos(-B/2J_x)$

the gap vanishes.

For $J_1 = J_2$, $J_3 = \lambda J_1 < 0$ (b) corresponds to the phase transition line $\lambda < 1$, $T_c > 0$ in Fig. 3.16, whereas (c) corresponds to the frustration point $\lambda = 1$, $T = 0$. For $\lambda < 1$ the excitation spectrum $\varepsilon(q)$ is <u>linear</u> around $q = \pi$ for $T = T_c$; for $\lambda = 1$ it is <u>linear</u> around $q = \pi/3$ for $T = 0$. Examples of $\varepsilon(q)$ from ref. 36 are shown in Fig. 3.19.

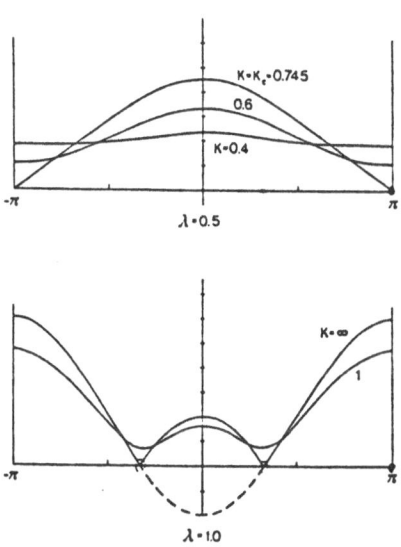

Fig. 3.19 Fermion excitation spectrum $\varepsilon(q)$ of the XY-chain corresponding to the AF triangular lattice with $\lambda = 0.5$ and $\lambda = 1.0$ (ref. 36).

Because of the linear behavior of $\varepsilon(q)$ in both cases one can construct continuous Luttinger models[48], which reproduce the asymptotic pair correlation function. For $\lambda < 1$ and $T = T_c$ Peschel obtains the power law behavior with the well known normal Ising exponent $\eta = 1/4$, whereas for $\lambda = 1$ and $T = 0$ he gets

$$ G^0(x) = \cos\left(\frac{2\pi}{3}\frac{x}{a}\right) \cdot \left(\frac{x}{a}\right)^{-1/2} \quad , \tag{3.44} $$

where a is the lattice constant. At the frustration point this method reproduces the value $\eta = 1/2$ and also in continuum version $(x/a = r)$ the oscillating prefactor first obtained by Stephenson (Eq. (3.33)).

The advantage of the present method is its applicability to the Union Jack and to Villain's odd model too. For both systems at the frustration point one obtains[36]

$$G(x) = \frac{1}{2}\left[1 + \cos \pi \left(\frac{x}{a}\right)\right] \cdot \left(\frac{x}{a}\right)^{-1/2} \quad , \tag{3.45}$$

that is, $\eta = 1/2$ here too; because of the prefactor correlation between spins with odd distance vanishes, just as for the lattice result[13]. We shall have a closer look at this result later.

A closely related problem, which can also be mapped on a 1d quantum spin chain Peschel and Emery[49] have considered to obtain exact correlation functions. This is the 1d kinetic Ising model introduced by Glauber[50] with the energy

$$H(\sigma) = -J \sum_n \sigma_n \sigma_{n+1} \quad , \tag{3.46}$$

where σ_n are classical Ising spins. Its time development is described by a master equation for the probability distribution $p(\sigma,t)$ [49,50]:

$$\frac{\partial}{\partial t} \tilde{p}(\sigma,t) = -\sum_{\sigma'} T(\sigma,\sigma') \, \tilde{p}(\sigma',t)$$

with

$$\tilde{p}(\sigma,t) = p_o(\sigma)^{-1/2} \cdot p(\sigma,t) \quad ; \tag{3.47}$$

$$p_o(\sigma) = \text{const. exp} \, (-\beta H(\sigma)) \quad .$$

$p_o(\sigma)$ is the equilibrium distribution. The time development operator T can be expressed by Pauli matrices and is hermitian[51]:

$$T = -\sum_{n=1}^{N} \left(A\sigma_n^x + B\sigma_{n-1}^z \, \sigma_n^x \, \sigma_{n+1}^z + C\sigma_n^z\sigma_{n+1}^z - D\sigma_n^z \, \sigma_{n+2}^z - E\right) \quad , \tag{3.48}$$

where the constants A, ..., E are functions of the three independent spin flip rates and of the nn-interaction J . This operator is, apart from the last two terms, identical to the Hamiltonian $\hat{H}$, Eq. (3.38), and in general (for $D \neq 0$) after a Wigner-Jordan transformation yields a system of interacting fermions.

The essential idea now is, that in a special subspace of the parameter
space of a 2d system (interaction constants and temperature) the
Hamiltonian H respective $\hat{H}$ (Eqs. (3.36b), (3.36a)) commuting with
its transfer matrix V becomes identical to T . Then the equilibrium
distribution function $p_o(\sigma)$ of the 1d kinetic Ising chain is also
an eigenfunction of the transfer matrix V of the 2d system, and
especially that one belonging to the largest eigenvalue. Therefore, in
this subspace the correlation functions of the 2d system can be obtained exactly and have strict 1d character (for the simple kinetic
Ising chain with ferromagnetic nn-interactions (Eq. (3.46)) G(r) decayes exponentially for all T > 0).

With this method Peschel and Emery[49] have exactly reproduced the disorder line of the AF triangular lattice, found already by Stephenson[13]. The advantage of this method is its flexibility, thus Peschel
and Emery also found a disorder line in the 2d uniaxial ANNNI-model
(however, only in the Hamiltonian limit, Eq. (3.36b)), which questions
the original phase diagramm. The detailed discussion of the ANNNI-model
follows in Section 3.4.1. A careful description of disorder and order
lines in systems with competing interactions can be found in Rujan
(refs. 52, 53).

The essential properties of the frustrated triangular lattice are
summed up now and the corresponding parameter range is shown in
Fig. 3.20 (which differs from Fig. 3.12 because the isotropic AF
point $AF_o = -(3)^{-1/2} \cdot (1,1,1)$ is now in the center and the ferromagnetic points F_n at the corners.

Apart from the general case (a) , $J_1 \neq J_2 \neq J_3$ there are two special
cases, $0 > J_n = J_{n+1} > J_{n+2}$ (b) , which are the double lines from $K_{\bar{n}}$
to AF_o , and finally the central point AF_o itself (c) .

(a) A normal Ising transition occurs at a finite T_c into a LRO
     phase. At a temperature $T_D > T_c$, the disorder temperature of
     Stephenson, the pair correlation function changes its asymptotic
     behavior: whereas for $T < T_D$ the wavevector is fixed (q = 0
     or q = π) , for $T > T_D$ it changes <u>continuously</u>. According to
     Stephenson this is characteristic of a disorder surface of the
     first kind in the parameter space of the system. The disorder sur-
     face does not correspond to some phase transition; S, U and c_v
     have no singularity. For $T = T_c$ the correlation functions show
     power law decay, $r^{-\eta}$, with the usual 2d Ising value $\eta = 1/4$.

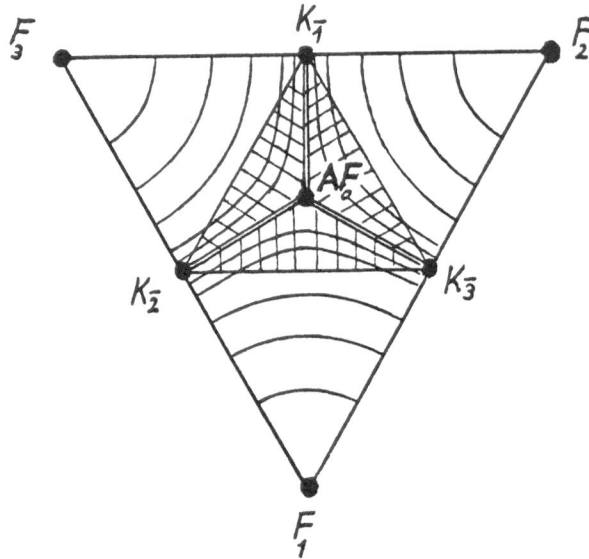

Fig. 3.20 AF triangular lattice. Lines of $T_C/(J_1+J_2+J_3)$ = const. . In the hatched
area the system is frustrated. Along the double lines from K_π to AF_O
T_C vanishes.

(b) For T = 0 only in the direction of the strong interaction there
is 1d LRO , as in single chains. Therefore, the GS entropy
S_O = 0 . In the other two nn-directions the correlation function
decays as $r^{-1/2} \cos(\pi r/2)$. Thus chains with odd distance are
uncorrelated, and for the ones with even distance η = 1/2 . This
difference in η to the usual Ising value shows that the frustrated system belongs to a different universality class. As
$G(r) \propto r^{-1/2}$ for T = 0 , the correlation length diverges and
one calls this a phase transition with T_c = 0 .

(c) For isotropic interactions (point AF_O) the frustration is strong
enough to inhibit any LRO at T = 0 , and the GS entropy becomes finite ($S_O \simeq 0.323$) . In general, however, these two properties are independent. For T > 0 the system is paramagnetic,
$G(r) \propto \exp(-r/\xi(T))$. For T = 0 in all three nn-directions
$G(r) \propto r^{-1/2}$; thus again η = 1/2 and T_c = 0 as in case (b) ,
but now the system is isotropic.

3.3 Further Frustrated Systems With Noncrossing Interactions

After the detailled discussion of the AF triangular lattice in the
last section (3.2), we may describe the other lattices of Table 3.1,2
more briefly. Here we consider only those models with noncrossing pair
interactions. Then the structure of the 1d operator of the transfer
matrix is not essentially more difficult than for the triangular lattice and an exact solution is possible. Along the same lines also periodically stacked models can be solved which are translational invariant only in one direction; in the perpendicular one the interaction constants are periodically repeated after ν lattice constants[54].

The anisotropic triangular lattice, and also some of the following
systems can be considered as special cases of these horizontally or
diagonally stacked systems, to which we turn at the end of this section.

3.3.1 Union Jack Lattice

As the triangular lattice with nn-interactions the Union Jack lattice
can be regarded as a square lattice where apart from nn-interactions
J_1 there is one diagonal interaction J_2 in each elementary square
as shown in Fig. 3.21.

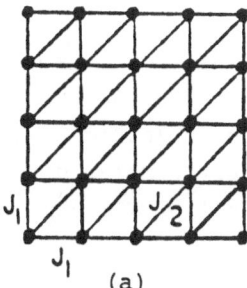 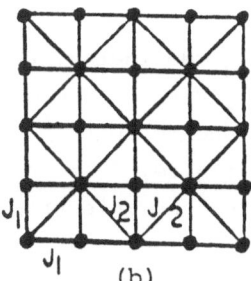

Fig. 3.21 Comparison of the triangular lattice (a) and the Union Jack lattice (b).

In this lattice there are two nonequivalent square sublattices. The
spins of sublattice 1 (S1) interact via J_2, the spins on sublattice
2 (S2) interact via J_1 with the spins of S1 only. For $J_2 < 0$ (AF)
independent of the sign of J_1 (here assumed to be positive) each
elementary triangle is frustrated. This model was first investigated
by Vaks, Larkin and Ovchinnikov 1966 [55], who obtained exact results for
the free energy and for the pair correlation function of sublattice
S1 .

Let us first consider the GS as a function of the ratio of the inter-
actions $\lambda = |J_2|/J_1$. Like in the triangular lattice there are three
different cases $\lambda < 1$, $\lambda = 1$ and $\lambda > 1$.

$\underline{\lambda < 1}$: The interactions J_2 are weak, the GS has ferromagnetic
LRO , is only twofold degenerate, and $S_o = 0$.

$\underline{\lambda = 1}$: The point $(\lambda = 1, T = 0)$ is the frustration point of the
Union Jack lattice. In each elementary triangle any one inter-
action may be broken. This leads to a GS degeneracy $N_g \propto c^N$
and thus to $S_o > 0$. The degeneracy N_g is equal to the
number of differently covering its dual 4-8-lattice (see
Tab. 3.2b) with nn-dimers, for which a simple lower bound can
be found:

$$N_g > 2^{N/2} (1 + 1/16)^{N/2} = (17/8)^{N/2} .$$

From this follows

$$S_o > \frac{1}{2} \ln \frac{17}{8} . \tag{3.49}$$

$\underline{\lambda > 1}$: Now J_2 dominates, the spins on S1 for $T = 0$ have complete AF LRO . Therefore, the spins on S2 are completely
free, and in spite of LRO on one sublattice the GS entropy is finite, $S_o = 1/2 \ln 2$. A very similar situation we shall
find for the anisotropic 'odd model' of Villain.

We add that $S_o(\lambda=1)$ must be larger than $S_o(\lambda>1)$ because for $\lambda > 1$
only interactions J_1 are broken, whereas for $\lambda = 1$ both J_1 or J_2
interactions can be broken.

Figure 3.22 shows the phase diagramm of Vaks et al.[55] together with the
disorder line T_D (Stephenson[13]), where as abscissa and ordinate

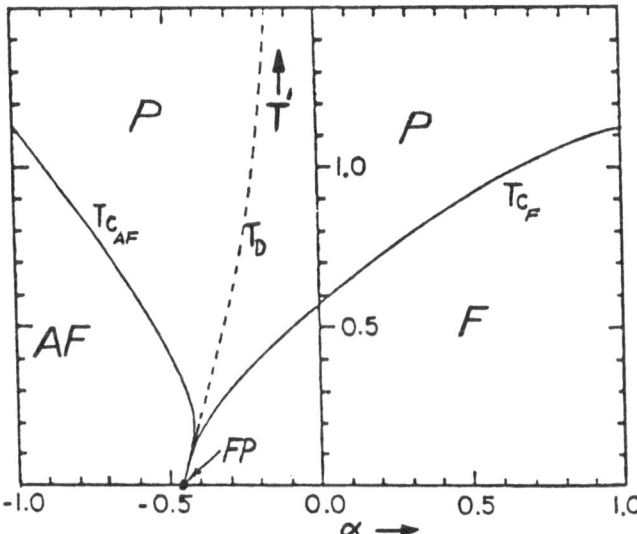

Fig. 3.22 Phase diagramm of the Union Jack lattice. The full lines are the transition temperatures T_{CF} and T_{CAF} of the F and AF phases. The dashed
line is the disorder line T_D .

the scaled quantities $\alpha = - \lambda/\sqrt{4 + \lambda^2}$ and $T' = K_1^{-1}/(2\sqrt{4+\lambda^2})$ are
used.

At the frustration point $(\lambda = 1, T = 0)$ $\alpha_{FP} = - 1/\sqrt{5}$. For $\alpha > \alpha_{FP}$ and
$\alpha < \alpha_{FP}$ the F and AF groundstates are connected to corresponding
$(T > 0)$-LRO phases. Between the transition lines T_{CF} and T_{CAF} the
paramagnetic (P) phase extends down to the frustration point, in
agreement with the exponential decay of the diagonal correlation function along the disorder line[13]

$$\cosh 4K_1^D = e^{-4K_2^D} \tag{3.50}$$

for $T > 0$:

$$G(r) \propto \begin{cases} \left(\dfrac{\pi r}{2}\right)^{-1/2} (\tanh K_2)^r & \text{for even } r \\[2em] 0 & \text{for odd } r \ . \end{cases} \tag{3.51}$$

For $T_D \ll J_1$ from Eq. (3.50) one obtains the linear relation

$$T_D/J_1 \equiv (K_1^D)^{-1} = \frac{4}{\ln 2} (1-\lambda) \quad , \quad \lambda \leq 1 \quad . \qquad (3.52)$$

As for $T \ll J_1$ the F and AF phase approach the disorder line exponentially, in a range

$$0.907 \stackrel{<}{\sim} \lambda < 1 \qquad (3.53)$$

the system shows reentrent behavior. With decreasing temperature three consecutive transitions occur:

1.) paramagnetic - antiferromagnetic,

2.) antiferromagnetic - paramagnetic,

3.) paramagnetic - ferromagnetic.

At all three transitions asymptotically

$$G(r) \propto r^{-1/4} \quad , \qquad (3.54)$$

that is, all transitions belong to the universality class of the non-frustrated Ising model. However, this is not true for the frustration point, which we consider in more detail. Such an unusual multiple phase transition is not only interesting theoretically, but also found experimentally for rochelle salt[56]. The multiple transition can be understood as an effect of the competing interactions leading to an effective nn-interaction of sublattice S1 which changes sign at some finite temperature[30]. Fradkin and Eggarter[57] have investigated decorated Ising systems which are similar to the Union Jack lattice and also show multiple transitions. In Fig. 3.23 to the right of the unit cell of the Union Jack lattice the one of the simplest decorated system is shown, where the interior spins can be summed out leading to an effective interaction

$$K_{eff} = K_2 + \frac{1}{2} \ln (\cosh 2K_1) \qquad (3.55)$$

(compare Eq. (3.9)) with a schematical temperature dependence shown in Fig. 3.24 for fixed $\lambda = - K_2/K_1$ [30] .

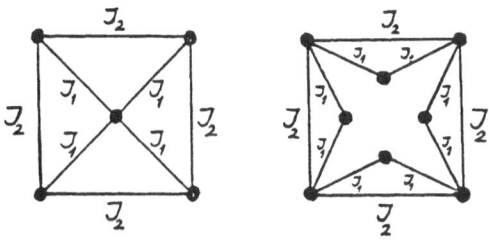

Fig. 3.23   Comparison of the unit cell of the Union Jack lattice with a model of
            Fradkin and Eggarter.

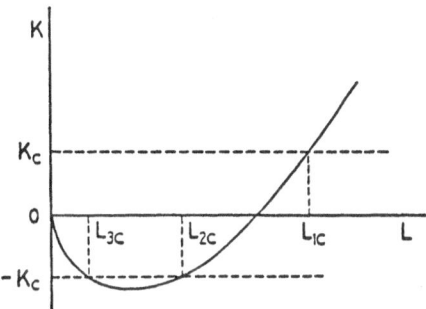

Fig. 3.24   Effective temperature dependent interaction K as function of L = J₁/T
            (ref. 30).

For convenient λ at low temperatures $(L > L_{1C})$ $K_{eff} > + K_c$, where
K_c is the inverse transition temperature of the F square Ising lattice, and the system is in the F phase. With increasing temperature
at L_{1C} the system becomes paramagnetic, at L_{2C} it becomes AF
ordered again. Finally when $|K|$ decreases below K_c at L_{3C} , there
is a third transition to the paramagnetic phase. This sequence of
transitions is identical to that of the Union Jack lattice in the
interval of λ given by Eq. (3.53).

The difference between the decorated model and the Union Jack lattice
where summation over the central spins of the unit cell (Star-square

transformation, Fig. 3.3c) generates more complicated interactions between the remaining spins, is obviously only quantitative. The occurrence of a multiple phase transition can be understood from the temperature dependence of the effective nn-interaction. The low temperature behavior of the specific heat near the frustration point as found by Vaks and Geilikman[43] is shown in Fig. 3.25.

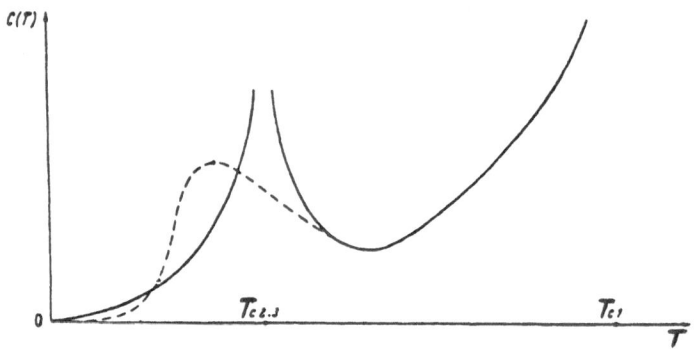

Fig. 3.25 Specific heat at the Union Jack lattice for $\lambda = 1 - \delta$ (with $|\delta| \ll 1$). For $\delta > 0$ (full line) two very close transitions occur at $T_{c2,3}$ corresponding to L_{1C} and L_{2C} of Fig. 3.24, in addition to a transition at higher temperature T_{c1}, existing also for $\delta < 0$ (dashed line) (after ref. 43).

For $\lambda = 1$ ($\delta = 0$) $c_v(T)$ is monotonous below T_{c1},

$$c_v(T) \simeq 6.4 \ K_1 \ e^{-4K_1} \ . \tag{3.56}$$

For $0 < 1 - \lambda \ll 1$ the two low temperature transitions L_{1C}, L_{2C} almost coincide with $(T_D/J_1)^{-1} = K_{D_1}$ and in Fig. 3.25 only one logarithmic divergence is visible. For $\lambda > 1$ there is no transition at low temperature, and similar to the triangular lattice only an analytical maximum near $T/J_1 = - (4/\ln 2) \ (1-\lambda)$ occurs.

Stephenson[13] has determined the diagonal pair correlation function $G(r)$ also away from the disorder line T_D (Eq. (3.50)). In the paramagnetic phase below T_D $G(r)$ decays monotonously $(q = 0)$, whereas above T_D $G(r) \cdot (-1)^r$ behaves like this $(q = \pi)$. The wavevector jumps at T_D from one fixed value to another one; the

Union Jack lattice exhibits a disorder line of the second kind[13].

Finally we consider $G(r)$ right at the frustration point. Extrapolation of Eq. (3.51) along the disorder line down to $T = 0$ yields:

$$G(r) \propto \begin{cases} r^{-1/2} & r \text{ even} \quad , \\ \\ 0 & r \text{ odd} \quad , \end{cases} \qquad (3.57)$$

thus $\eta = 1/2$ as in the isotropic AF triangular lattice.

Forgacs[25] used the exact results of Vaks et al.[55] to discuss in more detail $G(r)$ when the frustration point is approached from different directions. As can be seen from Eq. (3.51), approaching along the disorder line there is a crossover from exponential to $r^{-1/2}$ decay. Approaching the frustration point along the phase boundaries a similar crossover from $r^{-1/4}$ for $T > 0$ to $r^{-1/2}$ for $T = 0$ is found.

Both limits of $G(r)$ for $T \to 0$ are idential, as the two transition lines and the disorder line are exponentially close for $T = 0$; that is, Eq. (3.51) is valid for all three lines.

When approaching the frustration point in Fig. 3.22 from a different angle, one finds asymptotically:

$$G(r) = \text{const.} \qquad \text{(from the F phase)} \quad ,$$
and $\qquad (3.58)$
$$G(r) = (-1)^r \cdot \text{const.} \quad \text{(from the AF phase)} \quad ,$$

with const. ≤ 1 .

Therefore, the frustration point is a singular point, where the behavior of $G(r)$ depends on the direction of approach.

The mapping on the quantum XY-chain discussed for the triangular lattice is possible here too; Peschel[36] this way reproduced the $r^{-1/2}$ power law when approaching the frustration point along T_D .

3.3.2 Villain's Odd Model and its Generalizations

Villain[58] was the first to consider the fully frustrated Ising system on the square lattice with all nn-interactions of equal absolute value, where the number of AF bonds around each elementary square must be odd ('odd-model'). In Table 3.2c,d two domino models introduced by André et al.[59] are shown, which have one AF interaction J' and three F interactions J per elementary square and which are equivalent to Villain's odd model for $\lambda \equiv - J'/J = 1$. They differ only by the arrangement of the unit cells consisting of two squares with J' in the center (dominos) and behave differently for $\lambda \neq 1$.

3.3.2a Groundstate and Phase Diagrams

Let us first consider the PUD model (Piled-up domino) of Table 3.2c. Here the AF interaction J' form contiguous chains.

For $\lambda < 1$ the GS is F , the weak J' interactions are broken, and $S_o(\lambda < 1) = 0$.

The point $(\lambda = 1, T = 0)$ is the frustration point of the model with a high GS degeneracy identical to the number of complete dimer coverings of the dual square lattice. This equivalence yields the GS entropy per site[60]

$$S_o(\lambda=1) = \frac{G}{\pi} \simeq 0.2916 \qquad , \qquad (3.59)$$

where G is Catalan's constant.

For $\lambda > 1$ J' dominates, the J' chains for T = 0 are AF ordered. The total GS degeneracy can be obtained from the number of GS of the J chains which have ordered J' chains on both sides. The calculation is completely analog to the 1d ANNNI-chain, Section 2.1.1, and yields

$$S_o(\lambda>1) = \frac{1}{2} \ln \left(\frac{1+\sqrt{5}}{2} \right) \simeq 0.2406 \qquad . \qquad (3.60)$$

It is interesting to note the coupling of neighboring J' chains via an entropy effect. When the spins on both chains are oriented the same way, the intermediate J chain has a finite GS entropy per site.

However, for opposite spin orientation on the J' chains the GS entropy of the intermediate chain vanishes.

The PUD model has two phase boundaries, both ending at the frustration point (Fig. 3.26). The two boundaries of the F and AF phase are given by[59]

$$\sinh 2K \sinh (K+K') = 1$$

and   (3.61)

$$\sinh 2K \sinh (K+K') = -1 \quad .$$

Like in the Union Jack lattice also in the PUD model for $\lambda > 1$ there exists LRO on one sublattice, while the disorder on the other sublattice causes the entropy to remain finite down to $T = 0$.

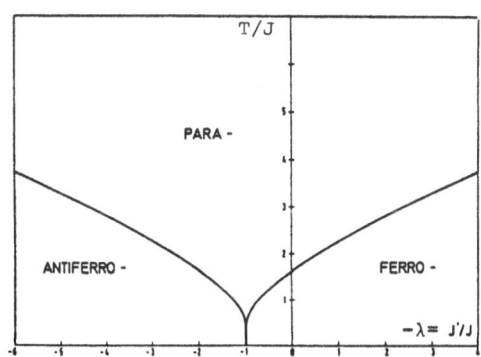

Fig. 3.26 Phase diagram of the PUD model (ref. 59).

The GS of the ZZD model (zick-zack domino), see Table 3.2d, is F ordered for $\lambda < 1$ too, with $S_o = 0$.

For $\lambda = 1$ like the PUD model it is equivalent to Villian's odd model, ($\lambda = 1$, $T = 0$) is its frustration point too.

For $\lambda > 1$ it differs markedly from the PUD model, as the strong J' interactions do not form contiguous chains. In the GS the pairs of spins interacting via J' are therefore coupled rigidly. As these pairs form a fully frustrated triangular lattice, the GS entropy

per pair is identical to that of the AF triangular lattice; S_o per
site thus is

$$S_o(\lambda > 1) = \frac{1}{2} S_o^{AF\Delta} \simeq 0.1615 \qquad .$$
(3.62)

The ZZD model possesses only an Ising transition for $\lambda < 1$ at[59]

$$2 \tanh (K+K') \tanh 2K = 1 \qquad .$$
(3.63)

For $\lambda \geq 1$ there is no singularity in the free energy for $T \neq 0$,
the system remains in the paramagnetic phase. The corresponding phase
diagram[59] is shown in Fig. 3.27.

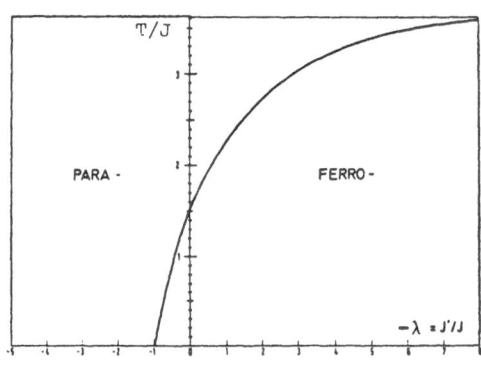

Fig. 3.27 Phase diagram of the ZZD model (ref. 59).

3.3.2b Correlation Functions

Consider the correlation functions for $T = 0$:

For $\lambda < 1$ PUD and ZZD models are ferromagnetic; $G^o(r) = 1$. For
$\lambda = 1$ (odd model) Forgacs[35], Gabay[61] and Peschel[30] have determined
$G^o(r)$ and all found $\eta = 1/2$. Wolff and Zittartz[54] discuss in detail
the behavior of $G(r)$ in horizontal, diagonal and vertical direction
for zero and finite temperature. Here we only mention the $T = 0$ results (without changing signs depending on the exact position of the
chain):

$$G^o(r) \quad \propto \quad \begin{cases} \text{const.} \cdot r^{-1/2} & ; \quad r \quad \text{even} \\ \\ \text{const.}/\sqrt{2} \cdot r^{-1/2} & ; \quad r \quad \text{odd} \end{cases} \qquad (3.64a)$$

for horizontal and vertical directions, and

$$G^o(r) \quad \propto \quad \cos\left(\frac{\pi}{2} r\right) \cdot r^{-1/2} \qquad (3.64b)$$

for diagonal direction.

For $\lambda > 1$ and $T = 0$ the J' chains of the PUD model are ordered, whereas the intermediate J chains cause the finite GS entropy. The correlation functions along the J' and J chains are ((b) is only valid for $r \to \infty$):

$$G^{o'}_{PUD}(r) \quad = \quad (-1)^r \quad , \qquad (3.65a)$$

$$G^o_{PUD}(r) \quad = \quad \frac{1}{5} (-1)^r \quad . \qquad (3.65b)$$

For $\lambda > 1$ and $T = 0$ apart from the entropy also $G^o(r)$ of the ZZD model in vertical and diagonal direction is equivalent to the AF triangular result:

$$G^o_{ZZD}(r) \quad \propto \quad r^{-1/2} \cos\left(\frac{2\pi}{3} r\right) \quad . \qquad (3.66)$$

3.3.2c Periodical Layered Models

The PUD and the ZZD models can be considered as the simplest vertically (or horizontally) layered models, where the interactions are arranged translation invariant within each layer and the layer repetition period is $\nu = 2$. Wolff, Hoever and Zittartz in a series of papers[54,62] have investigated general layered models and have calculated their phase diagrams and correlation functions. The method and the results are summarized in ref. 54.

We want to discuss two more of these models. The first one, a generalization of the PUD model with $\nu = 2$, is an example clearly demonstrating how little GS properties are related to the properties at
the transition, if $T_c > 0$. Calling the J' interactions of the PUD
model J_1 and modifying the J interactions parallel to the J_1
ones to become J_2, for $T = 0$ one obtains a 'phase diagram'[63]
(Fig. 3.28) where only along the thick lines $S_o > 0$. The spin configuration changes along these lines and also along the dashed part of

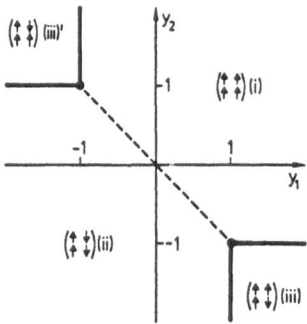

Fig. 3.28 Groundstate phase diagram of the modified PUD model (ref. 63).

the diagonal $y_1 + y_2 = 0$ in Fig. 3.28, where $y_i = J_i/J$. In
Fig. 3.29 the GS energy E_o is plotted as a function of y_1 and
y_2. But only along the whole diagonal $y_1 + y_2 = 0$ one has $T_c = 0$,

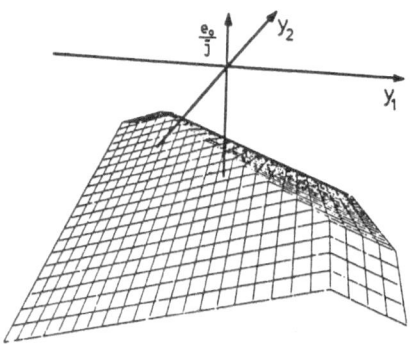

Fig. 3.29 The corresponding groundstate energy (ref. 63).

everywhere else a normal Ising transition occurs at a finite T_c depending only on $|y_1 + y_2|$ (see Fig. 3.30).

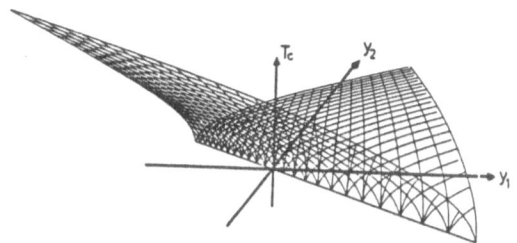

<u>Fig. 3.30</u> The corresponding transition temperature T_c (ref. 63).

The $T = 0$ 'phase diagram' indicating changes of the GS spin configurations does not have any clear influence on T_c. It is interesting to consider the horizontal and vertical correlation function for $T_c = 0$, that is for $y_1 = - y_2 > 1$ (if $y_1 < 1$, only the horizontal and vertical $G(r)$ are exchanged).

Along the horizontal direction

$$G_{h_1}(r) = 1 \qquad , \qquad G_{h_2}(r) = (-1)^r \qquad , \qquad (3.67)$$

whereas along the vertical direction similar to the anisotropic frustrated triangular lattice for $r \to \infty$ [58]

$$G_v(r) \quad \propto \quad \left| \cos\left(\frac{\pi}{2} r\right) \right| \cdot r^{-1/2} \quad ; \qquad (3.68)$$

for odd distance r the horizontal chains are also completely decoupled, and for even r again $\eta = 1/2$.

The frequent occurrence of the value $\eta = 1/2$ in the examples of frustrated systems discussed up to now which have no finite T_c had led to the assumption[35] that all such systems become critical at $T_c = 0$ with $\eta = 1/2$. However, we are now going to discuss as a first counter example the chessboard model with different behavior.

3.3.2d Chessboard Model

The chessboard model[59,54,64] is a diagonal layered model with period
$\nu = 4$ shown in Table 3.2e where every second elementary square is
frustrated. We mention this model as the first one with a finite correlation length down to $T = 0$, and thus not becoming critical at
$T = 0$.

Already André et al.[59] found the absence of a transition for $T > 0$
and also the existence of a finite GS entropy S_O . Wolff and Zittartz[64] have obtained:

$$S_O \simeq 0.371 \qquad . \qquad\qquad\qquad (3.69)$$

In ref. 54 they show that the diagonal correlation function decays exponentially for $T = 0$ with the short correlation length ξ_O :

$$\xi_O^{-1} = \ln (1 + \sqrt{2}) \qquad . \qquad\qquad\qquad (3.70)$$

As the reason for this behavior different from the previous systems
with $T_c = 0$, Süto[65] considers 'superfrustration', for which he de-
rives criteria. He distinguishes between frustrated systems with respective without 'isolated' GS , that is GS , which cannot be reached
from other GS by flipping locally only a limited number of spins. He
conjectures systems with isolated GS to have long range correlations
and thus $r^{-\eta}$ decay, whereas systems without isolated GS should have
only short range correlations and $e^{-\xi/r}$ decay.

We now turn to three other models also exhibiting exponential decay
for $T = 0$ which demonstrate the chessboard model to be no peculiar
case.

3.3.3 Hexagon Lattice

The hexagon lattice (Tab. 3.1b) can be regarded as a special case of a diagonal layered square lattice where in every second row every second interaction is omitted as shown in Fig. 3.31.

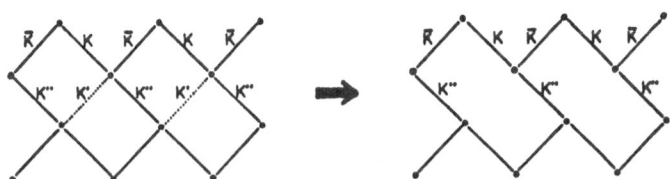

Fig. 3.31   Transformation of a square lattice into a hexagon lattice. The dotted interactions have to be omitted (ref. 66).

This way Wolff and Zittartz[66] have solved the general anisotropic and thus also the fully frustrated Ising system on the hexagon lattice with the configuration of interactions $K_i = \pm K$ shown in Fig. 3.32.

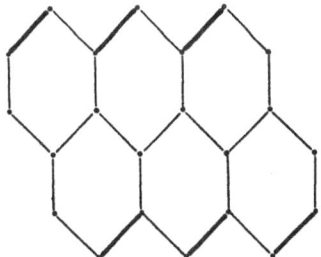

Fig. 3.32   Configuration of interactions of the fully frustrated hexagon lattice. Thick lines are AF , thin lines F interactions (ref. 66).

The system has a finite GS entropy[66]

$$S_o \simeq 0.214 \quad ,$$

(3.71)

and is paramagnetic for all temperatures down to T = 0 , where the correlation length remains finite[66]

$$\xi_0^{-1} \;=\; \ln\,(2+\sqrt{3}) \qquad . \qquad\qquad (3.72)$$

As the chessboard model the fully frustrated hexagon lattice does not
become critical at T = 0 .

3.3.4 Pentagon Lattice

The ferromagnetic and the fully frustrated AF pentagon lattice have
been investigated by Waldor, Wolff and Zittartz[67], as another example
of a layered system which is shown in Fig. 3.33.

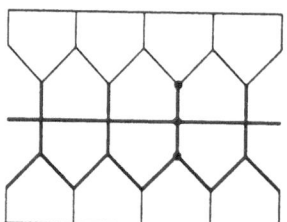

Fig. 3.33 Pentagon lattice; one layer is drawn with thick lines (ref. 67).

Whereas in the F case they found the usual Ising behavior as expected,
in the AF case the system has a finite GS entropy[67]

$$S_0 \;\approx\; 0.2336 \qquad\qquad (3.73)$$

and is paramagnetic down to T = 0 . The temperature dependence of the
horizontal correlation length $\xi_h(T)$ and of the wavevector $q = \theta(T)$
of the asymptotic oscillations of $G_h(r)$ are shown in Fig. 3.34.

Like the systems of the two previous sections the AF pentagon lattice
also does not become critical at T = 0 .

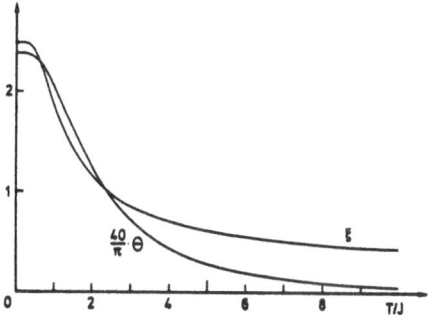

Fig. 3.34 Horizontal correlation length ξ_h and wavevector Θ of the AF pentagon
lattice as a function of temperature.

3.3.5 Kagomé Lattice

As for the triangular and the square lattice the properties of the Kagomé lattice (Table 3.1c) have been studied for different configurations of the interactions K_i .

The simplest case with frustration is the isotropic AF case, where
all interactions are equal. Then all elementary triangles in the lattice are frustrated, but not the hexagons. Kano and Naya[68] have found
the absence of a singularity in the free energy for $T > 0$; therefore, the system is paramagnetic in this range. The GS entropy is
finite and very high:

$$S_o \simeq 0.5018 \quad . \tag{3.74}$$

A very similar value is obtained from the Pauling approximation mentioned in connection with the triangular lattice, which treats the
triangles as being independent:

$$S_p^P = \ln 2 + \frac{2}{3} \ln \frac{3}{4} \simeq 0.5014 \quad . \tag{3.75}$$

The reason for this good agreement is probably the very weak pair correlation which we shall discuss further.

Figure 3.35 shows the internal energy of the Kagomé lattice with F
respective AF interactions together with the corresponding triangular
lattice results[68], which for T > 0 look similar.

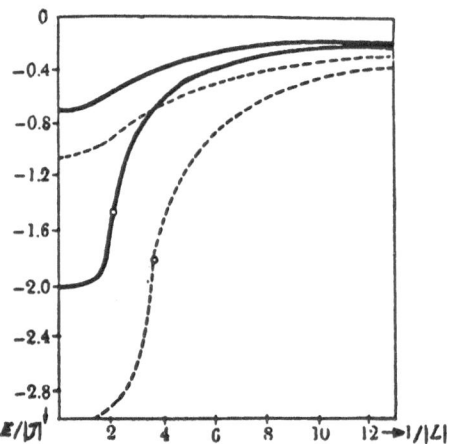

Fig. 3.35 Temperature dependence of the internal energy of the AF (upper) and
F (lower) Kagomé lattice (full lines). The corresponding triangular
lattice results are shown as dashed lines (ref. 68).

Although Kano and Naya have calculated the partition function of the
anisotropic Kagomé lattice (with three different interactions K_i for
the three different nn-directions), only Geilikman[69] for the case
$J_1 = J_2 = J > 0$, $J_3 = - \lambda |J| < 0$ shown in Fig. 3.36a has discussed
the phase diagram and the correlation functions along the dashed lines
in Fig. 3.36a. The sign of J is unimportant in the absence of a
magnetic field.

For weak J_3 ($\lambda < 1$) he finds a normal Ising transition to a simple
F phase, and the transition temperature for $\lambda \to 1$ vanishes linear:

$$K_c^{-1} = \frac{T_c}{J} \simeq \frac{1}{\ln 2} (1-\lambda) \qquad . \qquad (3.76)$$

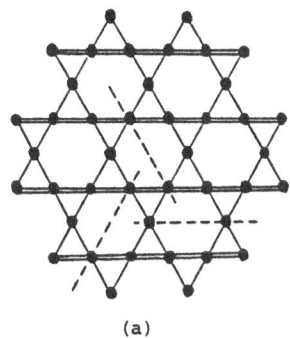 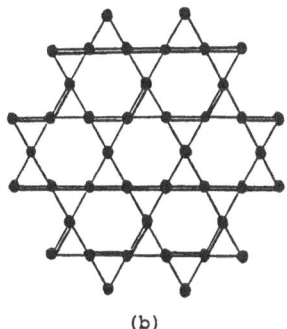

(a)   (b)

Fig. 3.36 Frustrated Kagomé lattice; the double lines correspond to AF interactions J_3 .

(a) Only the triangles are frustrated, along the dashed lines the pair correlation function is discussed in ref. 69;

(b) here the hexagons are frustrated too.

The correlation function $G(r)$ along the dashed lines in Fig. 3.36a vanishes completely along a disorder line T_D ($T_D > T_c$) which is also linear for $\lambda \to 1$ [69]:

$$K_D^{-1} = \frac{T_D}{J} \simeq \frac{2}{\ln 2} (1-\lambda) \qquad .\qquad\qquad (3.77)$$

The point ($\lambda = 1$, T = 0) is the frustration point like for the AF triangular lattice (Fig. 3.16).

For $\lambda > 1$ the J_3 chains order for $T \to 0$ but no order occurs in the perpendicular direction (according to ref. 69 for $T \to 0$ $\xi_t \to 0$; for T = 0 all three correlation functions $G_i(r)$ vanish).

For $\lambda > 1$ the system is paramagnetic for finite temperature. Süto[70] has investigated for a whole family of frustrated Kagomé models with different configurations of F and AF interactions the behavior of the free energy and the pair correlation function for $T \geq 0$. In these models apart from all triangles also an arbitrary part of the hexagons is frustrated.

To the right side of Fig. 3.36a, where no hexagon is frustrated, Fig. 3.36b shows one possible realization of the other extreme where all hexagons are frustrated. Süto shows that independent of the frustration of the hexagons the free energy outside the hatched area of the complexe tanh (β |J|) - plane in Fig. 3.37 is analytic including the whole positive real axes. This means there can be no singularity in the free energy at any temperature including T = 0 . For the correlation function he obtains for T ≥ 0 the upper bound[70]

$$|\langle s_i s_j \rangle| \quad < \quad 4 \cdot (0.74)^{|i-j|} \quad , \quad (3.78)$$

corresponding to a maximal correlation length

$$\xi_{max}^{-1} \quad \gtrsim \quad 0.30 \quad . \quad (3.79)$$

In the proof he expands $\langle s_i s_j \rangle$ in powers of the nn-pair correlation function ζ_{nn} of two spins on a single frustrated triangle

$$|\zeta_{nn}| \quad = \quad \frac{w}{1 + w + w^2} \quad = \quad \frac{1 - x}{3 + x} \leq \frac{1}{3} \quad , \quad (3.80)$$

with $w = |\tanh K|$ and $x = e^{-2|K|}$. This power series converges very fast; the simplest approximation for the internal energy (that is, the nn-correlation function)

$$U = -2 |\zeta_{nn}| \quad (3.81)$$

agrees already to within a few percent with the exact U .

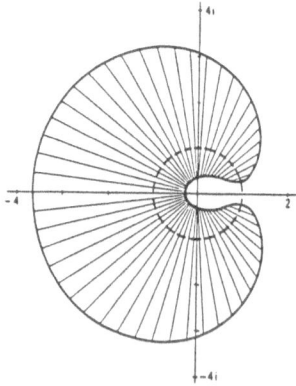

Fig. 3.37 In the non-hatched area of the complexe tanh (β |J|) - plane the free energy of all Kagomé models with fully frustrated triangles is analytic.

Equation (3.78) means exponential decay of the correlation function also for $T = 0$. Thus the frustrated Kagomé lattice is the forth system besides the chessboard and the frustrated pentagon and hexagon lattices, which does not become critical at $T = 0$.

The exact internal energy U and the entropy S are known only for the pure AF Kagomé lattice where no hexagons are frustrated. As Eq. (3.81) is a good approximation also in this case, $U(T)$ and $S(T)$ can depend only weakly on the frustration of the hexagons.

We also note that integrating $U(T)$ from Eq. (3.81) exactly yields the Pauling approximation for S_0, Eq. (3.75), which also was very close to the exact result.

3.3.6 Connection Between GS Degeneracy and Existence of a Phase Transition at $T_C = 0$

At the end of Sections 3.2 and 3.3 where we have discussed 2d frustrated Ising systems solved exactly, we want to mention more general considerations on the connection between the degeneracy of the ground-state and the existence of a transition.

Hoever, Wolff and Zittartz[71] have formulated the following conjecture:

> If the set of all GS is connected, that is if any two GS can be transformed into each other by a series of purely local transformations, the global symmetry of the Hamiltonian, $s_i \rightarrow - s_i$, cannot be broken. In this case there is no phase transition.

In case of the chessboard[71] and the AF Kagomé lattice all GS are connected by 1-spin flip processes; in the hexagon lattice always 2 nn spins must be flipped simultaneously to obtain one GS from another one. For all three systems $\xi(T=0)$ remains finite, these systems do not become critical at $T = 0$ in agreement with the above conjecture.

If the GS are not connected, there are no general statements; Hoever et al.[71] mention examples with and without a transition.

Süto has also put foreward a conjecture consistent with the three systems just mentioned:

If and only if a frustrated Ising system also for T = 0 has
no transition and ξ(T=0) remains finite, the set of all GS
is connected.

Süto calls such models 'superfrustrated'. Both conjectures still have
to be proven.

3.4   Frustrated Ising Systems With Crossing Interactions

In this section we discuss the   2d ANNNI-model and the triangular and
the square lattice with nn and crossing nnn interactions which,
therefore, cannot be solved exactly.

For comparison with the ANNNI-model we also consider the (simplified)
brick model solved exactly and finally comment on the connection of
these models to vertex models.

3.4.1   2d   ANNNI-Model

The 2d ANNNI-model is derived from the   1d ANNNI-chain (compare
Sec. 2.1) by repetition in one perpendicular direction and introducing
additional nn-interactions J_0 > 0 between spins on neighboring
chains; the corresponding lattice is shown in Table 3.2i.

The GS are analog to the   1d case: F   for $\lambda = - J_2/J_1 < 0.5$
(J_1 > 0) , and periodic ↑↑↓↓ = <2> for λ > 0.5 ; at the frustration
point 1d disorder occurs.

The essential difference to the   1d chain is the existence of transitions with $T_c \neq 0$. Of the well known methods to determine phase
diagrams the low temperature series (does not converge for d = 2)
and mean field approximation (too large fluctuations) cannot be used.
Thus the phase diagram of Fig. 3.38 is combined from Monte Carlo (MC)
calculations[12,72] for not too low T , the Müller-Hartmann/Zittartz

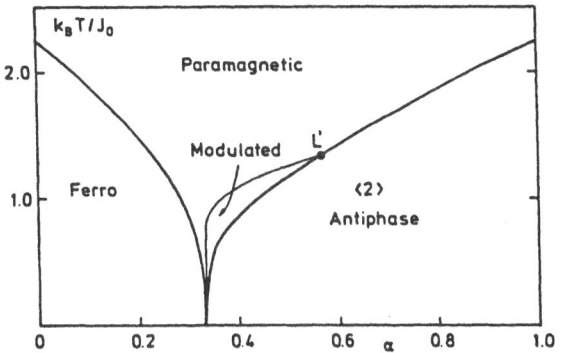

Fig. 3.38 Phase diagram of the 2d ANNNI-model with $J_1 = (1-\alpha) J_0$ and $J_2 = - \alpha J_0$ (ref. 72).

approximation for domain boundary energies[12,73,76] and a free fermion approximation[74] good for $T \ll J_0$, $|J_2|$.

Adjacent to the LRO groundstate for $\lambda < 0.5$ respective $\lambda > 0.5$ are finite temperature phases with the same order. Not common and, therefore, especially interesting is the occurence of an incommensurate modulated M phase between the <2> and the P phase; the exact location of the Lifshitz point L'[75], a special multicritical point, is not yet known.

With the method of Müller-Hartmann and Zittartz the phase boundaries can be estimated analytically[12,73a]:

$$F : \quad \sinh 2(K_1+2K_2) \, \sin 2K_0 \; = \; 1 \qquad , \qquad (3.82)$$

and

$$<2> : \quad \exp (2K_0) \; = \; (1 - \exp (4K_2))/\{(1 - \exp (-K_1+2K_2)) \cdot$$
$$(1 - \exp (K_1+2K_2))\} \qquad . \qquad (3.83)$$

They are in good agreement with the MC data[72].

The MC data also indicated first the existence of the incommensurate modulated M phase by another maximum of the specific heat.

Contrary to the original assumption[12] of a Lifshitz point occuring at
(λ < 0.5, T > 0) where F , M and P phase all are in equilibrium,
now there are many indications for the P phase to reach down to the
frustration point (λ = 0.5, T = 0) between the F and M phase.

The free fermion approximation (FFA) of Villain and Bak[74] is very
helpful to understand the behavior of the M phase. They start from
the assumption supported by the MC data, that in the M phase at
low temperature only such spin configurations occur, which can be
described by a set of domain walls with a minimal distance r = 2 .
Thus the walls do not touch each other, but don't have to be straight
as for T = 0 .

Figure 3.39 shows one such domain wall configuration.

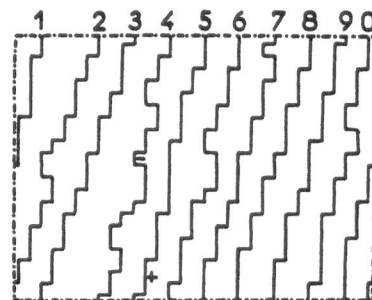

Fig. 3.39 A typical domain wall configuration included in FFA (ref. 74).

The summation over these 'dislocation free' wall configurations can
by done analytically and leads to a free fermion problem where the
largest eigenvalue is connected to the average distance between walls
and thus determines the wavevector q .

As visible in Fig. 3.40, q increases continuously from the F to
the <2> phase boundary. For the transitions Villain and Bak[74] obtain

$$F : K_1 + 2K_2 = e^{-2K_0} \qquad (3.84a)$$

and

$$<2> : K_1 + 2K_2 = -2 e^{-2K_0} \qquad ; \qquad (3.84b)$$

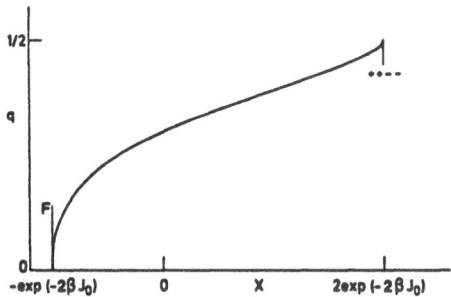

Fig. 3.40 Wavevector q of the modulated phase as a function of
x = - $(J_1+2J_2)/T$ (ref. 74).

the first one corresponds to Eq. (2.82) for K_o >> 1 , whereas the
second one differs by the factor two on the right hand site from the
(K_o >> 1) expansion of Eq. (3.83).

The FFA result for the pair correlation function G(r) in the direction of the competing interactions is[74]:

$$G(r) \simeq r^{-\eta} \cos \pi q x \qquad (3.85a)$$

with η depending continuously on temperature and λ :

$$\eta = \frac{1}{2} (1-q)^2 \qquad . \qquad (3.85b)$$

Within the framework of the FFA which assumes nontouching domain
walls, the phase boundary between M and P phase cannot be investigated.

The inclusion of a low concentration of dislocations (that is of de-
fects where walls touch) corresponds to taking care of the vertices
in the 2d XY ferromagnet, for which Kosterlitz and Thouless[77] have
also found a modulated phase with

$$G(r) \propto r^{-\eta} \qquad (3.86)$$

where $\eta(T)$ is temperature dependent. The equivalence of the M
phases in the two models together with the known T_c of the XY-model

yields the transition temperature between the M and P phases of the ANNNI-model[74]. For

$$q < 1 - \frac{1}{\sqrt{2}} \qquad\qquad (3.87)$$

the M phase cannot be stable. Therefore, the P phase must reach down to T = 0 in a narrow region between the F and M phases.

This result is consistent with the disorder line T_D found by Peschel and Emery[49] in the 'Hamiltonian limit' ($J_o \to \infty$, $J_1 \to 0$, $J_1/J_2 =$ const.) which extends down to T = 0 and approaches the F transition line exponentially. Along T_D G(r) decayes exponentially, that is T_D is in the P range of the phase diagram.

Using a transfer matrix approach Pesch and Kroemer[73b] have studied the ANNNI-system for semiinfinite strips (N × ∞; N ≤ 13) . They determine entropy and specific heat as functions of N and T . The latter result is quite close to their results in the Hamiltonian limit, supporting the conclusions derived from this approximation. In the correlation function across the finite width of the strips they can see the transition from ferromagnetic to oscillating behavior although because of the finite width the dependence of the wavevector on J_2/J_1 can only be estimated. The critical exponents β/ν and η obtained from their finite-size analysis deviate from the Ising values for $|J_2|/J_1 \gtrsim 0.3$.

This leaves the open question, whether the deviation is due to slower convergence of the finite size analysis because of the proximity of the para-to-modulated transition in case the paramagnetic phase indeed extends down to T = 0 , or whether there is a modified transition.

The 2d ANNNI-model thus is a relatively simple Ising system where the competing interactions lead to an incommensurate modulated phase with continuously varying wavevector and short range order (SRO) . In the other systems discussed up to now below T_c always LRO existed, at least on a sublattice.

In the chapter on 3d systems we shall see that the 3d ANNNI-model differs markedly from the 2d case near the frustration point.

3.4.2 Brick Model

This model can be derived from the 2d ANNNI-model by omitting alternately every second interaction J_o perpendicular to the direction of competing interactions $(J_o' \to 0)$ and doubling the strength of the remaining ones $(J_o'' \to 2 J_o)$, see Table 3.2h.

Then no crossing nnn-interactions are left and the system can be solved with transfer matrix methods as in Section 3.3 (Bidaux and de Seze[78]).

Although the difference between the ANNNI and the brick model may appear quite small, the second model for $\lambda \geq 0.5$ has a by far larger GS degeneracy. Thus for $\lambda = 0.5$ the GS entropy is finite and half the value of the AF triangular lattice[37]

$$S_o(\lambda = 0.5) \quad = \quad \tfrac{1}{2} \, S_o^{AF-\Delta} \quad \simeq \quad 0.16153 \quad . \tag{3.88}$$

For $\lambda > 0.5$ the GS is ordered only in the direction of the competing interactions.

Mean field approximation here as for the 2d ANNNI-model would lead to a completely wrong phase diagram, whereas the exact solution shows that only for $\lambda < 0.5$ there is a transition from an F to a P phase, with $T_c = \beta_c^{-1}$ given by[78]:

$$\tanh (2\beta_c J_o) \, \sinh (2\beta_c J_1) \, \exp (4\beta_c J_2) \quad = \quad 1 \quad . \tag{3.89}$$

The phase diagram of the brick model for different values of J_2/J_1 is shown in Fig. 3.41. For $\lambda > 0.5$ there is no transition at $T_c \neq 0$; and for $T = 0$ $G(r)$ in J_o direction vanishes exactly for $r \geq 2$.

Morgenstern[79] has considered a somewhat generalized brick model where the nn-interactions in axial direction are either J_1 or $J_3 > J_1$, causing the disappearance again of the 1d GS degeneracy for $\lambda = - J_2/J_1 > 0.5$ (also he takes $J_o = J_1$). From the exact solution for $\lambda > 0.5$ he obtains precisely one transition, whereas MC simulations of finite systems even after a very long time (about 120.000 possible spin flips per site) still exhibited modulated phases which can only be metastable.

This demonstrates the necessity to interprete MC calculations very carefully when stable and metastable incommensurate modulated phases

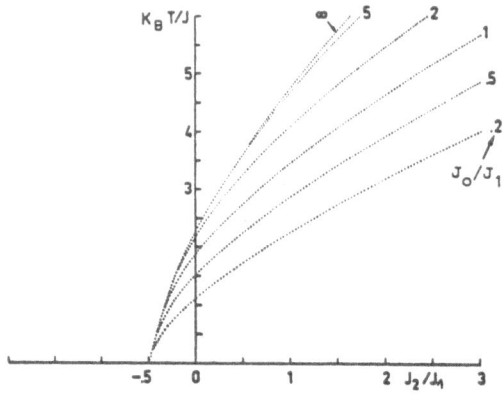

<u>Fig. 3.41</u> Phase diagram of the brick model for different values of J_0/J_1 as a
function of $-\lambda = J_2/J_1$ (ref. 78).

may occur. Thus in finite systems it may be difficult to distinguish
between phases where $G(r)$ decays with a power law and others with
LRO ; periodical boundary conditions can stabilize metastable states.

3.4.3 Frustrated Triangular Lattice With nnn-Interactions and Magnetic Field

In Section 3.2 we have discussed the frustrated triangular lattice with
(an-) isotropic nn-interactions J_1 . This model is of great theoretical interest, but there are hardly any corresponding experimental systems, because further interactions even if they are weak have a strong
influence e.g. on the GS degeneracy.

In this section we want to consider the model with additional nnn-
interactions J_2 and magnetic field H . This is not only interesting
for understanding magnetic quasi-2d systems like $ErGa_2$[80], but also
as the lattice gas model for the study of thin adsorption layers of
noble gases on graphite[81].

3.4.3a   Additional nnn-Interactions J_2

The GS of the triangular lattice with J_1 and J_2 interactions has been investigated by Metcalf[82], Tanaka and Uryu[83] and Kaburagi and Kanamori[84], who have found four different LRO phases shown in Fig. 3.42. The GS phase diagram (Fig. 3.43) shows the GS phase

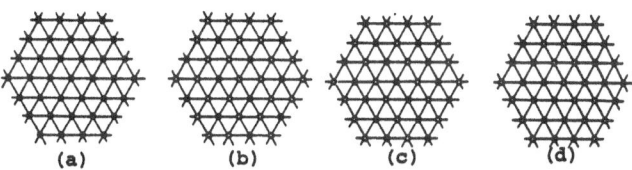

Fig. 3.42   The four LRO groundstates of the triangular lattice with nn- and nnn-interactions: (a) ferro, (b) $\sqrt{3} \times \sqrt{3}$, (c) 1×2 and (d) 1×4 (ref. 85).

boundaries. Whereas the GS degeneracy is finite in the interior of the four phases (2-, 6-, 6- and 12-fold for the ferro, $\sqrt{3} \times \sqrt{3}$, 1×2 and 1×4 phases), along the boundaries (apart form the one between the ferro and $\sqrt{3} \times \sqrt{3}$ phases) the degeneracy is large enough to give a finite GS entropy per site $S_0 > 0$.

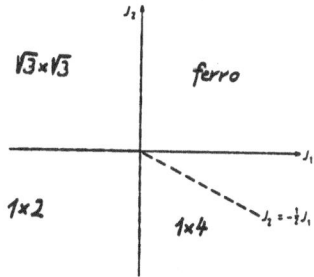

Fig. 3.43   GS phase diagram of the triangular lattice with nn- and nnn-interactions J_1 and J_2 .

An overview of the transitions shows Fig. 3.44. The T_c lines for the ferro and the $\sqrt{3} \times \sqrt{3}$ phases Oitmaa[85] has determined from high temperature series expansions of the corresponding order parameter susceptibilities, where in both cases he has assumed normal critical behavior (e.g. $\chi \propto ((T-T_c)/T_c)^{-\gamma}$) in the analysis. Whereas for the ferro phase he obtains a value γ close to the exact nn-value $\gamma = 1.75$, for the $\sqrt{3} \times \sqrt{3}$ phase with $\gamma = J_2/J_1 = -1$ he gets $\gamma \sim 2.4 \pm 0.002$ at $K_1^C = 0.505 \pm 0.005$. This value of γ is much higher than in the nn Ising or $(q = 3)$ Potts model.

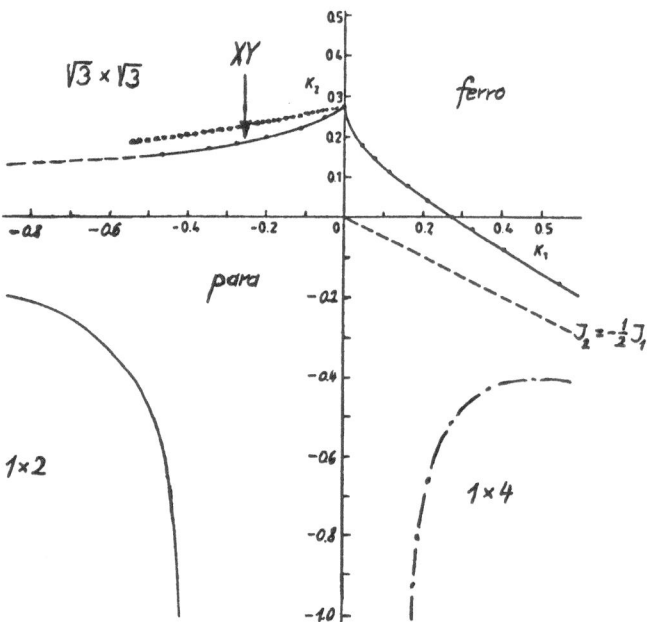

Fig. 3.44 Phase diagram of the triangular lattice with nn- and nnn-interactions.

This phase transition for $J_1 < 0$ and $J_2 > 0$ is of special interest because Domany et al.[86] have predicted that it belongs to the universality class of the XY-model with 6th order anisotropy. For this model an intermediate phase is expected between the ordered and the paramagnetic phases. Landau[88] has done extended finite size scaling analysis with MC calculations without and with a magnetic field H ; here we first discuss the H = 0 results.

In a large range of λ (-1/16 > λ > -8) the specific heat of systems
of size L × L with L = 30 exhibits two maxima as a function of
temperature (Fig. 3.45); the lower one is located very close to
T = 3.6 J_2 , independent of J_1 . At this temperature SRO occurs
within the three sublattices with interaction J_2 .

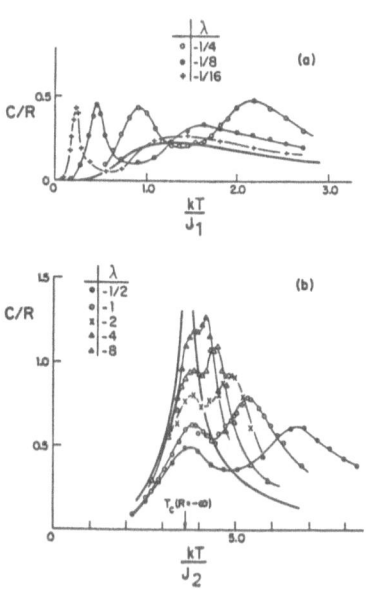

Fig. 3.45a,b Temperature dependence of the specific heat in a large range of
λ = J_2/J_1 .

(a) : $\lambda \geq -1/4$; (b) : $\lambda < -1/4$.

The abscissa in (a) and (b) have different scale (ref. 88).

Whereas Wada and Ishikawa[89] interprete this maximum as an indication
of divergencies in the infinite system, Landau has shown by finite
size analysis (for λ = -1) that this maximum remains finite for
L → ∞ , the exponent α thus must be negative and T_c cannot be
determined from the position of the maximum (Fig. 3.46).

Landau further finds that the correlation length ξ within the single
sublattice when approaching the transition from above and below

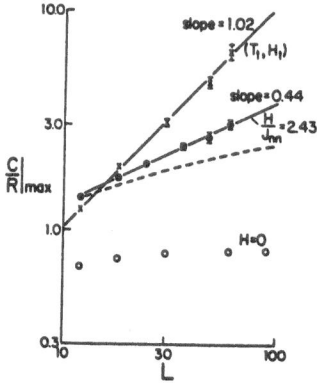

Fig. 3.46 Size dependence of the maximum of the specific heat for $\lambda = -1$ (ref. 88).

diverges exponentially at different temperatures $T_1 = 4.89\ J_1$ and $T_2 = 4.26\ J_1$:

$$\xi = \xi_o\ \exp\ a\ \left(\left| \frac{T-T_i}{J_1} \right| \right)^{-1/2} \qquad . \qquad (3.90)$$

Because of this exponential divergence (in contrast to the usual power law $\xi = \xi_o \cdot ((T-T_c)/T_c)^{-\nu})$ like in the XY-model a modified version of finite size scaling[88] must be used to analyse the order parameter susceptibility and the inverse order parameter. Both quantities scale very well when T_1 respective T_2 are inserted for T_c . Finally analysis of the pair correlation function shows power law behavior in the whole range $T_2 \leq T \leq T_1$: $G(r) \propto r^{-\eta}$, with η being continuously dependent on T , in good agreement with the theoretical predictions for the XY-model[77,87]:

$$\eta_{MC}(T_1) = 0.27 \pm 0.02 \qquad , \qquad \eta_{th}(T_1) = 1/4 \qquad ,$$

$$(3.91)$$

$$\eta_{MC}(T_2) = 0.15 \pm 0.02 \qquad , \qquad \eta_{th}(T_2) = 1/9 \qquad .$$

This confirms the interesting phase diagram expected for the frustrated triangular lattice with $H = 0$ and $\lambda < 0$:

$T < T_2$:   LRO   phase with finite sublattice magnetisation;

$$\lim_{r \to \infty} G(r) = const.(T) \quad .$$

$T_2 < T < T_1$:   XY   model-like phase without   LRO , the 'critical'
exponents are temperature dependent; in this range
a line of fixpoints occurs; $G(r) \propto r^{-\eta(T)}$.

$T > T_1$:   Paramagnetic phase; $G(r) \propto \exp(-r/\xi(T))$.

Landau has found this behavior for $\lambda = -1$. The dependence of T_1 and
T_2 on λ has not yet been investigated, but the intermediate XY-phase
certainly occurs in a larger range of λ . Therefore, in Fig. 3.44 in
addition to the transition line corresponding to the divergence of the
susceptibility at T_1 , a second transition line corresponding to T_2
has to be drawn; it is shown schematically by the dotted line.

The phase boundary of the 1×2 phase in Fig. 3.44 has been investigated up to now only by a mean field calculation[90] not leading to the
exact result $T_c = 0$ for $J_2 = 0$ and by Slotte and Hemmer[91]. These
authors used the Müller-Hartmann/Zittartz method for the calculation
of the domain wall energy[76]. Apart from the isotropic case they also
considered the cases where there are nnn-interactions only in two or
one direction. The transition line of Fig. 3.44 is drawn from their
tabulated data. Results for the order of the phase transition and values of the critical exponents do not yet exist.

The phase boundary of the 1×4 phase in Fig. 3.44 is purely schematic, no results are available.

3.4.3b Additional Magnetic Field H

Switching on a homogeneous magnetic field   H   further increases the
number of phases of the frustrated triangular lattice.

Figure 3.47 shows the   GS   phase diagram of Kaburagi and Kanamori[92]
where in addition to the four phases for $H = 0$ also a 2×2 and
a 3×3 respective 1×3 phase occurs. On the other hand for
$H \neq 0$ the transition between the ferro and paramagnetic phases dis-
appears.

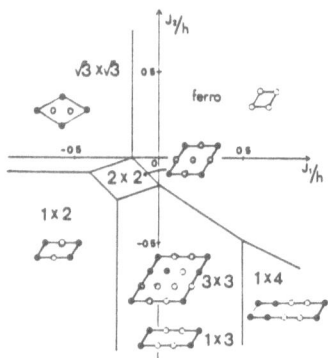

Fig. 3.47 GS phase diagram of the triangular lattice with nn- and nnn-interactions J_1 and J_2 in a magnetic field H (ref. 92).

Consider the $\sqrt{3} \times \sqrt{3}$ phase for $H > 0$. Then only three GS exist with spins on two sublattices parallel to H and on one antiparallel. For this case Alexander[93] and Domany et al.[86] have predicted the transition to belong to the universality class of the $q = 3$ Potts model. For $\lambda = 0$ ($J_2 = 0$) the magnetic field for $0 < H < H_{c_1} = 6 |J_1|$ leads to a transition at finite T_c. The $T_c(H)$ curves determined by Monte Carlo calculations[88], real space renormalization[94] and Müller-Hartmann/Zittartz method[95] agree qualitatively and are shown in Fig. 3.48a,b.

For $\lambda < 0$ ($J_2 > 0$) and $H = 0$ T_c is finite, the case $\lambda = -1$ has already been discussed. For $H > 0$ the phase transition is second order up a tricritical point $H = H_t$, for $H_t < H < H_c$ it is first order. The phase diagram determined by Landau[88] for $\lambda = -1$ is shown in Fig. 3.49 where the tricritical points are marked by crosses.

Figure 3.46 demonstrates the divergence of the specific heat at $H = 2.43 J_1$ with an exponent $\alpha/\nu \simeq 0.44$ (the 'best' Potts model value is $\alpha/\nu = 0.40$). Finite size scaling[40-42] yields: $\beta = 0.11$, $\gamma = 1.42$, $\nu = 0.87$ and $\eta = 0.27$, very close to the 'exact' Potts values: $\beta = 1/9$, $\gamma = 13/9 \simeq 1.444$, $\nu = 5/6 \simeq 0.867$ and $\eta \simeq 0.266$.

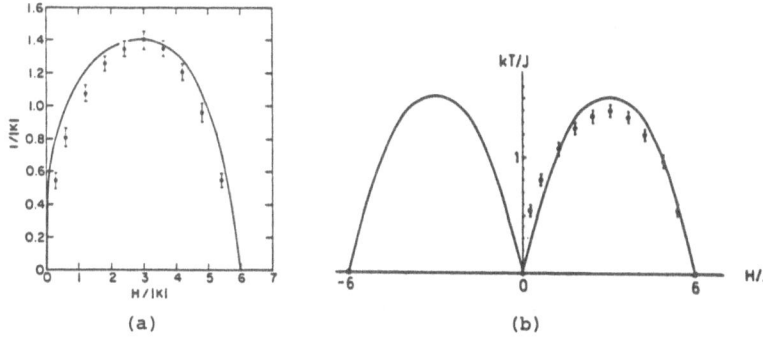

(a)   (b)

Fig. 3.48   Phase diagram of the AF triangular lattice in a magnetic field.
(a) RS-RG and MC results ; (b) MHZ and MC results (refs. 88,
94,95).

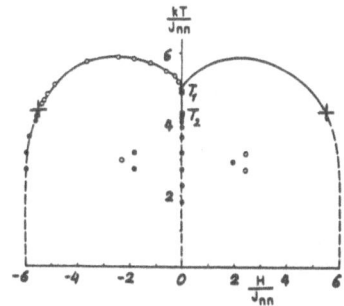

Fig. 3.49   Triangular lattice in a magnetic field for $J_2/J_1 = -1$. At $|H| = H_t$
tricritical points (crosses) occur where the order of the transition
changes from second (below) to first (above) (ref. 88).

The MC results are thus consistent with the prediction that the phase
transition for $0 < H < H_t$ belongs to the universality class of the
$q = 3$ Potts model.

Landau also examined the critical exponents at the tricritical point
and at the crossover from XY to $q = 3$ Potts behavior for small H ,
but this we do not discuss here.

A good experimental example for a frustrated Ising system on the triangular lattice with negative interactions J_1 and J_2 is $ErGa_2$, for
which Doukouré and Gignoux[80] have measured magnetisation and neutron
scattering. At low temperature they find the 1 × 2 phase for
H < H_{C_1} = 6.8 kOe , for H_{C_1} < H < H_{C_2} = 20 kOe the 2 × 2 phase and
finally for H > H_{C_2} the ferro/para phase (Fig. 3.50). This is exactly
the sequence of transitions expected for T = 0 in the range
1/4 < λ < 1 (see Fig. 3.47). At higher temperatures (above the transitions of the 1 × 2 and 2 × 2 phases) the magnetisation of $ErGa_2$
increases monotonous until saturation (Fig. 3.50).

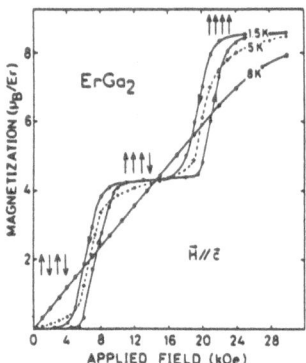

Fig. 3.50 Magnetisation M of $ErGa_2$ as a function of the magnetic field
          (ref. 80). For low temperatures two critical magnetic fields occur
          where M rises abruptly.

A more precise test for this system is not yet possible, because for
H ǂ 0 no phase diagram has been determined. For the vicinity of the
boundary between the 2 × 2 and 3 × 3 (or 1 × 3) phases in
Fig. 3.47 Nakanishi and Shiba[96] discuss modulated phases within mean
field approximation which, however, yields wrong results for 2d systems as has been proven for the ANNNI-case. We come back to this paper
in the chapter on 3d systems where this approximation should be qualitatively correct.

3.4.3c Corresponding Lattice Gas Model

A 2d Ising spin configuration can be taken as a representation of a
configuration of a submonolayer of adatoms on a surface: spins $s_i = +1$
represent occupied sites, $s_i = -1$ vacant ones. The AF nn-interaction J_1 causes repulsion between two adatoms on nn-sites, the magnetic field becomes the chemical potential determining the average coverage Θ ($\hat{=}$ magnetisation M). Such an adsorption model is called a lattice gas model.

For a more realistic description of the adsorption of noble gases like
Kr on hexagonal graphite layers (Fig. 3.51) one needs at least also
nnn-interactions as an approximation for the better Lennard-Jones po-
tentials[97].

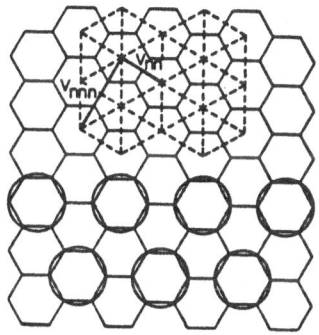

Fig. 3.51 Lattice gas model with nn- and nnn-interactions for the description of
adsorption of noble gas atoms on hexagonal graphite layers (ref. 97).

It may well be possible that still further reaching interactions are
necessary for the interpretation of adsorption layer measurements, but
these would lead to very complicated phase diagrams. Therefore, we
here refer to the paper by Kanamori[98] who has determined GS phase
diagrams for the hexagon lattice with first to third nn-interactions
and in special ranges of the parameters found "devil's staircase" behavior[99].

After the mapping on the lattice gas the Ising results can be reinter-
preted; one only has to note that experimentally in adsorption layers

the coverage Θ (≙ M) and not the chemical potential μ (≙ H) is the
independent variable. For λ = - 1 in the Θ - T phase diagram[88]
(Fig. 3.52) this leads to large coexistence regions.

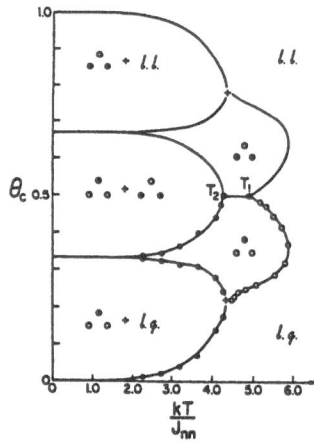 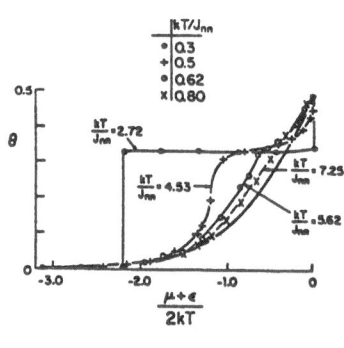

Fig. 3.52 Phase diagram in the coverage-
temperature phase for λ = - 1
(ref. 88).

Fig. 3.53 Adsorption isotherms for
λ = - 1 . The thick line
is Langmuir's isotherm
$(T \gg |J_{nn}|)$ (ref. 88).

The adsorption isotherms shown in Fig. 3.53 exhibit steps for $T < T_2$,
T_t corresponding to the successive covering of the three sublattices.
For $T > T_2$, T_t these steps become washed out and in the limit
$T \gg |J_{nn}|$ approach the Langmuir isotherm for noninteracting adatoms.

A review on theories and experiments of thin adsorption layers on solid
surfaces is given by Dash[100], the field is still in rapid development.

Further physical systems for which the Ising results can be applied are
binary alloys, where the atoms of species A and B can be represented
by s_i = + 1 respective s_i = - 1 . As alloys are usually 3d systems,
we postpone discussion to Chapter 4.

3.4.4 Square Lattice With Competing nn- and nnn-Interactions; Relation to Vertex Models

In this section we consider Ising models where frustration does not arise from competing nn-interactions as for example in Villain's odd model, but is caused by competition between nn- and nnn-interactions J_1 and J_2 in a magnetic field present or not.

First e.g. Dalton and Wood[101] concluded from series expansions for lattices with $J_1 > 0$ and $J_2 > 0$, that the critical exponents are completely independent of the ratio $\lambda = J_2/J_1$. However, this has turned out to be true only for one of the two critical surfaces of a more general model. On the other one nonuniversal critical exponents occur, and in an additional magnetic field tricritical points emerge.

The mapping on the 8-vertex (Baxter) respective the 16-vertex models is briefly discussed, as well as the connection between the GS degeneracies of the latter model and the 2d 6-vertex (ice) model. In addition we refer to two papers of Miyashita[102] and Fujiki et al.[103] who studied the effect of additional third and fourth nn-interactions in the fully frustrated nn AF triangular and Villain's odd model, especially with regard to the occurence of an XY like phase transition.

3.4.4a System Without Magnetic Field

The GS phase diagram of the square lattice Ising system with nn- and nnn-interactions shows three low temperature ranges (F, AF, SAF) corresponding to the 1×1, $\sqrt{2} \times \sqrt{2}$ and 1×2 structures[104]. In Fig. 3.54 the transition lines are shown schematically.

Because the order parameter of the SAF phase is two-dimensional, Krinsky and Mukamel[7] have predicted the corresponding transition to belong to the universality class of the XY-model with cubic anisotropy and thus nonuniversal critical exponents should occur.

When using real space renormalization group (RS-RG) methods[104-106] it turns out that the recursion formulas for the interactions between block spins generate additional four-spin interactions J_4. Therefore, the model has been studied in the enlarged parameter space (J_1, J_2, J_4). This space also contains the special case $J_1 = 0$

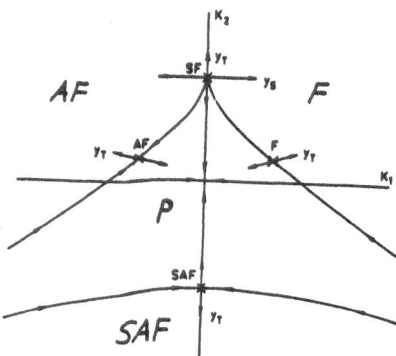

Fig. 3.54 Phase diagram of the square lattice with nn- and nnn-interactions (ref. 104).

which Kadanoff and Wegner[107] have proven to be equivalent to the 8-vertex model solved exactly by Baxter[108]. Thus the Ising model with J_2, $J_4 \neq 0$ is also called Baxter model.

The Baxter model has a second order transition with nonuniversal critical exponents depending continuously on the ratio: $\lambda = J_4/J_2$, indicating a whole line of fixpoints like in the 2d XY-model. Figure 3.55 shows the two sheets of the critical surface in (K_1, K_2, K_4) space.

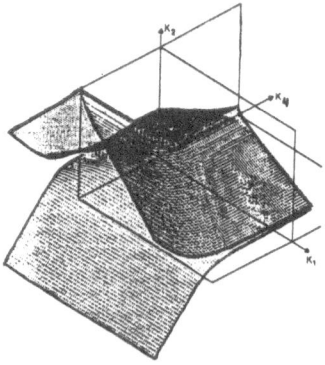

Fig. 3.55 The two sheets of the critical surface of the square lattice with nn , nnn and 4-spin interactions (ref. 105).

Figure 3.54 corresponds to the plane $K_4 = 0$ in Fig. 3.55. On both sheets there is a line of fixpoints at $K_1 = 0$ (Baxter case). In the upper sheet these fixpoints are repulsive, therefore, only exactly for $J_1 = 0$ one finds nonuniversal critical behavior for $J_4 \neq 0$. Contrary to this on the lower sheet the line of fixpoints is attractive. This is the reason why on the whole lower sheet, thus also for $J_4 = 0$ and $J_1 \neq 0$ nonuniversal behavior is found[106].

As an example in Fig. 3.56 the dependence of the exponent α of the specific heat is shown as function of $J_1/|J_2|$, as determined in perturbation theory by Barber[109]. For $J_1 = 0$ the system decomposes into two independent nn Ising square lattices. Then c diverges only logarithmically, and $\alpha(J_1=0) = 0$.

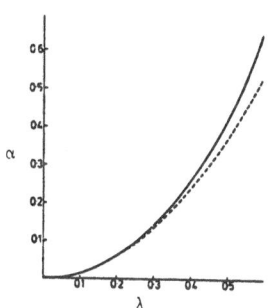

Fig. 3.56 Nonuniversal variation of the exponent α of the specific heat with $\lambda = J_1/|J_2|$ (ref. 109).

3.4.4b System With Magnetic Field

For finite magnetic field H and T = 0 an additional 1d degenerate 2 × 2 phase emerges between the F, AF and SAF phases (Fig. 3.57). The GS of this 2 × 2 phase exhibits perfect ferro order in every second chain (with spins parallel to the magnetic field) just as the SAF phase. However, the intermediate AF ordered chains are not correlated, whereas in the SAF phase the intermediate chains are F ordered (with spins antiparallel to the field).

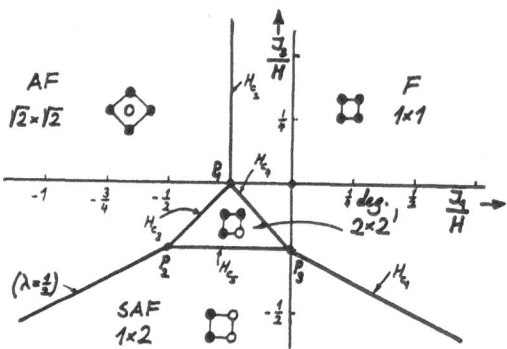

Fig. 3.57 GS phase diagram of the square lattice with nn- and nnn-interactions
          J_1 and J_2 in a magnetic field H .

The GS phase diagram (Fig. 3.57) apart from a correction of scale
is identical to the result of Kanamori[110a], who did not yet discuss
the degeneracy of the 2 × 2 phase. Brandt[110b] has extended this GS
phase diagram to third-nearest neighbors.

We now consider the phase transitions for AF nn-interactions
$(J_1 < 0)$, where the cases $\lambda = J_2/J_1 < 0$, $\lambda = 0$, $0 < \lambda < 1/2$,
$\lambda = 1/2$ and $\lambda > 1/2$ have to be distinguished.

In the first case $(\lambda < 0)$ the GS changes from AF to F at the
critical field $H_{c2} = 4 |J_1|$. The phase diagram for $\lambda = -1$ is
shown in Fig. 3.58 (ref. 111). For small H/J_1 there is an Ising

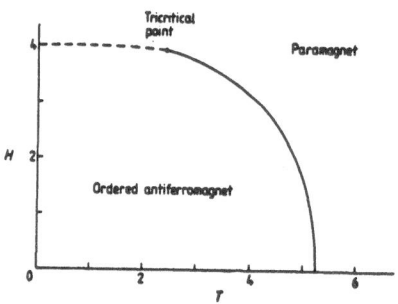

Fig. 3.58 Phase diagram of the square lattice with $\lambda = -1$ in a magnetic field
          (ref. 111).

transition with normal exponents (full line). Above the tricritical field H_t with modified critical exponents the transition becomes first order (dashed line). The interesting tricritical behavior has been treated by a number of authors[112].

For $\lambda \geq 0$ Binder and Landau[113] have carried out extensive MC calculations, and for $\lambda = 0$ (that is $J_2 = 0$) found excellent agreement with the results of Müller-Hartmann and Zittartz (MHZ) [76] which are based on the calculation of domain wall energies and at first were believed to be exact. We mentioned this method already in connection with the triangular lattice and the ANNNI-model. In the phase diagram (Fig. 3.59) also series expansion results[114] and RS-RG results[115] are included.

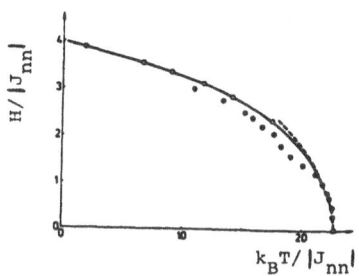

Fig. 3.59 Phase diagram of the nn AF square lattice in a magnetic field. Circles: MC results; full line: MHZ method; dashed line: series expansion; points: RS-RG (ref. 113).

In the meantime Baxter et al.[116] by high order series expansion have determined very precisely the critical activity z_c of the hard square lattice gas:

$$z_c = 3.7962 \pm 0.0001 \quad . \tag{3.92}$$

Contrary to this the MHZ result for the slope of the curve in Fig. 3.59, $m = d(H/J_1)/d(T/J_1)$ at the point ($H = 4 |J_1|$, $T = 0$) yields a value for the critical activity

$$z_c^{MHZ} = e^{-2m} = 4 \quad . \tag{3.93}$$

Therefore, the MHZ method cannot be exact, but it is a very good analytical approximation.

In the range $0 < \lambda < 1/2$ with increasing field H the GS changes at two critical fields H_{C_3} and H_{C_4} from AF to 2 × 2 to F = P phase as can be seen from Fig. 3.57. MC results of Binder and Landau[113] for $\lambda = 1/4$ (Fig. 3.60a) and the MHZ results of Doczi-Reger and Hemmer[117] (Fig. 3.61) agree quite well for the AF phase and show that along the line H_{C_3} the P phase reaches down to T = 0 in contrast to mean field approximation which yields a topologically wrong phase diagram.

For T > 0 there is no sharp transition between the 2 × 2 and the SAF phases as both have the same symmetry (as opposed to T = 0). Therefore, at T = 0 between H_{C_3} respective H_{C_5} and H_{C_4} an order-disorder transition occurs shown as thick line at T = 0 in Figs. 3.60a,b,c.

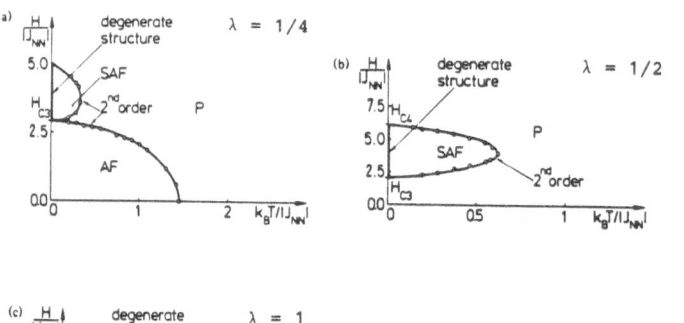

Fig. 3.60

Phase diagram of the square lattice with nn- and nnn-interactions in a magnetic field for (a) : $\lambda = 1/4$; (b) : $\lambda = 1/2$; (c) : $\lambda = 1$. All phase transitions are first order (ref. 113).

For $\lambda = 1/2$ (Fig. 3.60b) and $H < H_{C_3} = H_{C_5}$ the system remains paramagnetic down to T = 0 , between H_{C_3} and H_{C_4} the 2 × 2 phase (T = 0) and SAF phase (T > 0) occurs, with a second order transition to the P phase.

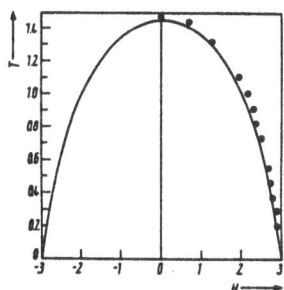

<u>Fig. 3.61</u> The AF phase transition in MHZ approximation (line) for $\lambda = 1/4$.
The points are the MC results from Fig. 3.60 (ref. 117).

For $\lambda > 1/2$ (Fig. 3.60c) already in low fields $H \geq 0$ the system is
in the SAF phase; for $T > 0$ only one contiguous transition line of
second order occurs, contrary to the behavior for $\lambda < 1/2$.

From finite size analysis Binder and Landau have found normal 2d
Ising exponents at the AF-P transition, whereas for the SAF-P
transition for $H \neq 0$ they find nonuniversal critical exponents in
agreement with the $H = 0$ results.

In the limits $1/\lambda \to 0$ and (for fixed λ) $H \to \infty$ the exponents approach the Ising values.

3.4.4c Connection to Vertex Models

Here we want to note the close connection between the Ising system
without field and the 8-vertex model and that between the system with
field and the 16-vertex model, which Lieb and Wu[118] have described extensively and which we mentioned already when discussing the Baxter
model.

Lieb and Wu show the equivalence of the Ising system with nn-interactions J_1 and J_2 , the nnn-interactions J and J' , the four-spin

interaction J and a constant J_o , with the 8-vertex model in which
the first eight vertices of Fig. 3.62 may occur. The energies e_i corresponding to the vertices are linear reversible functions of the Ising
interactions J_j .

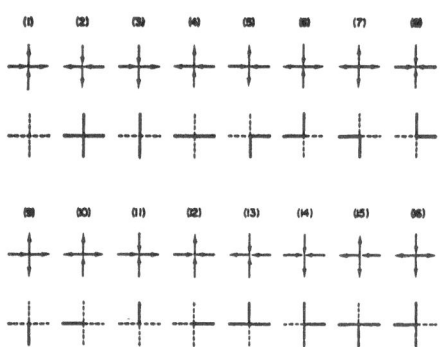

Fig. 3.62 The sixteen vertex configurations of the general ferroelectric model on
the square lattice and the corresponding bond configurations using vertex (1) as basis (ref. 118).

Lieb and Wu also show the Ising model shown in Table 3.2f, where nnn-
interactions occur only in every other square, to be equivalent to two
special cases of the 16-vertex model where all sixteen vertices of
Fig. 3.62 may occur. These special cases are called the generalized F
respective KDP model.

For both models the parameters can be chosen such that only the first
six vertices of Fig. 3.62 occur for T = 0 . The GS in these cases
corresponds to the square ice model[119,120], because these six vertices
must obey the 'ice-rule', which requires two of the four arrows at each
vertex to point towards it and the other two to point away from it[121].

In the equivalent Ising system of Table 3.2f all pair interactions are
AF and of equal strength. This system can be interpreted as a 2d
lattice of fully frustrated cornersharing tetrahedra (see Fig. 3.63).

For T = 0 in each tetrahedron two spins are up and two are down,
just corresponding to the ice rule. The entropy of the square ice model
Lieb has determined exactly[120]:

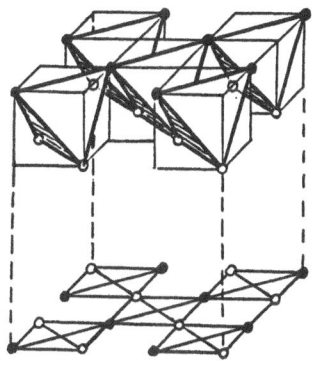

<u>Fig. 3.63</u> The square lattice (bottom) is equivalent to the 2d lattice of corner-
sharing tetrahedra (top).

$$S_o = \frac{3}{4} \ln \frac{4}{3} \approx 0.2158 \qquad , \qquad (3.94)$$

whereas the simple Pauling approximation[39] from the GS degeneracy of
single tetrahedra ($w_o^{tet} = 3/8$) yields the total degeneracy

$$N_g^P = 2^N \left(\frac{3}{8}\right)^{N/2} \qquad , \qquad (3.95)$$

and thus

$$S_o^P = \frac{1}{2} \ln \frac{3}{2} \simeq 0.2027 \qquad . \qquad (3.96)$$

We shall find the Pauling approximation to be even better for the isotropic 3d pyrochlore lattice in Section 4.3, which also consists of
cornersharing tetrahedra.

In general the frustration of an Ising model leading to multiple degeneracy of the GS shows up in the equivalent vertex model as the
existence of several vertices with equal minimal energy e_i^{min} .

Whereas in the ice model all six vertices have equal energy (maximal
frustration), in the F respective KDP model only two, and in the
inverse IF respective IKDP model only four vertices have lowest
energy. The reduction of the degeneracy of the vertex energies maps
into the reduction of pair interactions with equal strength in the

corresponding Ising model. When the horizontal respective vertical nn-
interactions in the lattice of Fig. 3.63 (bottom) are called J_x respective J_y and the diagonal nnn ones J , on obtains the Ising
GS phase diagram (Fig. 3.64), where the three LRO phases and se-
parating doublelines are named after the corresponding vertex model
phases[118]. Along the doublelines the (inverse) vertex models remain
paramagnetic down to $T = 0$; for $T \to 0$ only 1d LRO emerges.

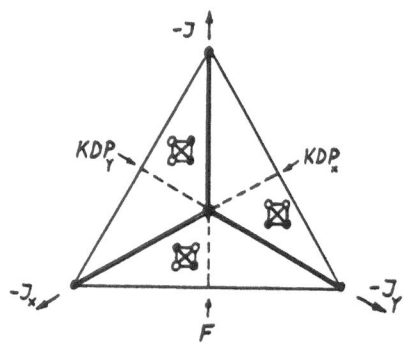

<u>Fig. 3.64</u> GS phase diagram of the Ising square lattice with crossing nnn-inter-
actions in every other square (Tab. 3.2f) for $J_x + J_y + J = \text{const.}$.

We cannot discuss further the interesting thermodynamic properties of
vertex models (e.g. phase transitions of infinite order and phase transitions of first order with exponentially diverging specific heat),
these are extensively discussed by Lieb and Wu. We only add that in
the KDP as well as in the F model for $T > T_c$ the pair correlation function $G(r)$ decays like $r^{-\eta}$. Both 6-vertex models which
obey the ice rule lie exactly on the critical surface of 8-vertex models which do no longer obey the ice rule[118].

Ising systems with general multi-spin interactions also can show
frustration effects leading to the vanishing of T_c at specific
frustration points. Here we mention only the triangular lattice with
nn-pair and three-spin interactions J_2 and J_3 in a magnetic field
H , which was solved exactly for $J_2 = H = 0$ by Baxter and Wu[122] and
in general has been studied by Doczi-Reger and Hemmer[123] using the
MHZ method. We shall discuss an example with four-spin interactions
for the fcc-lattice in the next chapter.

4. Three-Dimensional Frustrated Ising Systems

In the last chapter we have described a multitude of results for 2d frustrated Ising systems, many of them are exact. Contrary to this in three dimensions almost no exact results are known. An exception is the self-dual Ising system with four-spin interactions on the fcc- and hcp-lattice discussed in Section 4.5. First in Section 4.1 to Section 4.3 we treat three fully frustrated systems with nn-pair interactions on the fcc , the simple cubic (sc) and the pyrochlore (= B spinell) lattice, which show phase transitions of different order. In Section 4.4 the 3d ANNNI-model follows, which has properties differing markedly from the 2d case.

4.1 <u>fcc Antiferromagnet</u>

For this system with AF nn-interactions J_1 Danielian[124] has shown the GS to be only 2d ordered: perfectly AF ordered 100 planes can be stacked arbitrarily on each other. This yields a GS degeneracy

$$N_g \propto 2^{(N^{1/3})} \quad , \tag{4.1}$$

of course the GS entropy for $N \to \infty$ vanishes ($\propto N^{-2/3}$) . From low and high temperature series expansions Danielian gave a rough estimate of the transition temperature: $T_c \simeq 1.2 \ |J_1|$. Betts and Elliott[125] extended the low temperature series, but uncorrectly averaged over all (nonequivalent) GS and obtained $T_c \simeq (1.83\pm0.02) \ |J_1|$.

The interesting question how to do the low temperature expansion correctly has been investigated by several authors[126-128]. They have formulated the following conditions for the existence of such an expansion:

(a) Different GS for $N \to \infty$ must differ from each other by a diverging number of spins with different orientation.

$$\tag{4.2}$$

(b) The spin configurations of low energy excitations 'near' one GS may differ from the configuration of this GS for only a small (finite) number of sites.

$$\tag{4.3}$$

When these conditions are fulfilled one can determine the free energy $f_\alpha(T)$ starting from a GS (α) ; in general $f_\alpha(T)$ depends on α via the excitation spectrum.

Mackenzie and Young[127] have included a nnn-interaction J_2 and showed that for small J_2 only two sixfold respective twelfefold degenerate phases with maximum symmetry (and LRO) are stable thermodynamically. All other phases (with 1d disorder) have a higher free energy and thus can only be metastable. However, the energy differences are so small that these metastable phases in MC simulations at low temperatures will appear equally stable as the equilibrium phases.

Just as for several 2d systems LRO occurs at low, but finite temperatures, whereas for $T = 0$ there is 1d disorder.

Alexander and Pincus[8] have studied the GS properties and the Landau expansion of the free energy for generalized frustrated spin systems on d-dimensional fcc-lattices and spin dimension n . For $d = 3$ and $n = 1$ they expect an effectively 2d phase transition in the universality class of the 2d Heisenberg model with cubic anisotropy.

This phase transition between the ordered phase where two of the four sc-sublattices have spins up and the other two spins down, and the P phase has been investigated in a series of MC papers by Phani et al.[129] and Binder (et al.)[130-132], who also included a nnn-interaction J_2 and a magnetic field H .

For $H = J_2 = 0$ a first order phase transition occurs at $T/|J_1| = 1.75$. With increasing field H the GS changes at two critical fields $H_{c_1} = 4 |J_1|$ and $H_{c_2} = 12 |J_1|$ (see Fig. 4.1).

At these critical fields the GS entropy is finite[132] and the system remains paramagnetic down to $T = 0$:

$$S_0(H_{c_1}) \simeq S_0(H_{c_2}) \simeq \frac{1}{3} \ln 2 \qquad . \qquad (4.4)$$

Apart from these two points the phase transition is always first order. For $0 < H < H_{c_1}$ the ordered phase is threefold degenerate, for $H_{c_1} < H < H_{c_2}$ fourfold. Corresponding to these degeneracies, the phase transitions should be as in the $q = 3$ respective $q = 4$ Potts model[133].

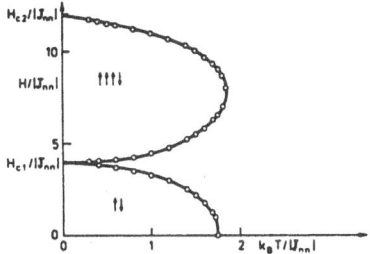

Fig. 4.1 MC phase diagram of the AF fcc-model in a magnetic field H
          (ref. 130).

Domany et al.[133] had predicted a new kind of critical behavior when
a nnn-interaction J_2 (> 0) is added. The two phase transition lines
come together at a multicritical point (T_m > 0, H = 0) . Exactly
this behavior Binder[132] has found for $J_2 = - J_1$ in MC calculations, although without determination of the critical exponents (see
Fig. 4.2).

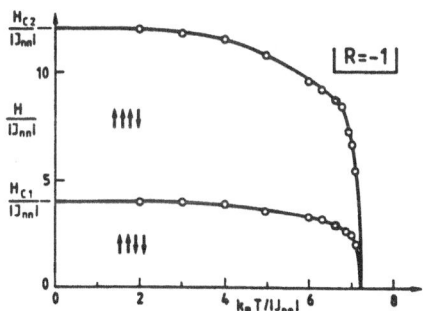

Fig. 4.2 MC phase diagram of the AF fcc-model with additional nnn-interactions
          J_2 (= - J_1) (ref. 132).

The fcc Ising system with J_1 < 0 and J_2 > 0 has been used to
describe $A_x B_{1-x}$ alloys. Correspondig results and further references
are contained in the MC papers cited here.

4.2 Fully and Partially Frustrated Simple Cubic Lattice

The properties of the fully and the partially frustrated simple cubic (sc) Ising system are not yet as well established as in the fcc-case. As square plaquettes are frustrated only for an odd number of AF interactions along the edges, one has to look for simple periodic configurations of F and AF nn-interactions on the sc-lattice leading to frustration of either (a) all plaquettes, or (b) only those oriented in the XY-plane. In Fig. 4.3 such configurations are shown. The unit cell contains four respective two elementary cubes.

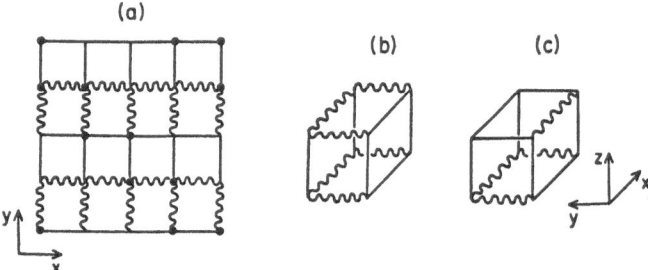

Fig. 4.3 (a) Comb representation of the fully frustrated square lattice. Straight (wavy) lines represent F (AF) interactions. Elementary cubes of the partially (b) and the fully (c) frustrated sc-lattice are obtained by stacking the square lattice of (a) in different ways (ref. 135).

Derrida et al.[134] have determined the GS properties of the fully frustrated system for d-dimensional hypercubic lattices and found that because of topological reasons for d > 4 in part of the plaquettes more than one interaction must be in the energetically unfavourable state, just as for d > 3 or 4 in the fcc-lattice[8]. This strong kind of frustration effect they have called 'superblocking'.

The GS degeneracy is higher then in the AF fcc-lattice (and in the frustrated diamond lattice[58]). Instead of 1d disorder one finds 2d disorder[136]: There is a set of GS where every fourth chain of spins parallel to one 100 direction can be flipped as a whole to get another GS from the previous one, leading to

$$S_o \propto N^{-1/3} \quad (sc) \quad , \qquad (4.5a)$$

in contrast to

$$S_o \propto N^{-2/3} \qquad \text{(fcc, diamond)} \qquad . \qquad (4.5b)$$

This has an important consequence for the low excited states.

When only a finite part of length L of such a chain is flipped from one GS configuration, additional energy arises only at the ends of this part; that is the difference

$$\Delta E = 4 |J| \qquad (4.6)$$

of these 'one-dimensional' excitations is independent of the length L . But this way condition (4.3) is violated, and no low temperature series expansion exists.

MC calculations of Bhanot and Creutz[137] and of Kirkpatrick[136] indicate a second order transition, but the critical temperatures are not in good agreement:

$$T_c/|J| \simeq 0.8 \qquad \text{(ref. 137)} \qquad ,$$
$$\qquad (4.7)$$
$$T_c/|J| \simeq 1.25 \qquad \text{(ref. 136)} \qquad ,$$

where additional data for structure factor and specific heat (Fig. 4.4) are in favour of the higher value.

Chui et al.[138] have tried to determine T_c from a mean field approximation using eight sublattices, but (contrary to their claim) they obtain a much higher value

$$T_c^{MF}/|J| \simeq 2.4 \qquad \text{(ref. 138)} \qquad . \qquad (4.8)$$

This demonstrates that systems with strong frustration and corresponding large fluctuations are not well described by mean field theory.

Blankschtein et al.[135] have determined the Ginzburg-Wilson Hamiltonian for the fully and a partially frustrated sc-system (see Fig. 4.3) and predict the following behavior:

(a) Fully frustrated:

The order parameter has four components $(n = 4)$; ε $(\equiv 4-d)$-RG expansion up to ε^2 is consistent with a transition of (weak)

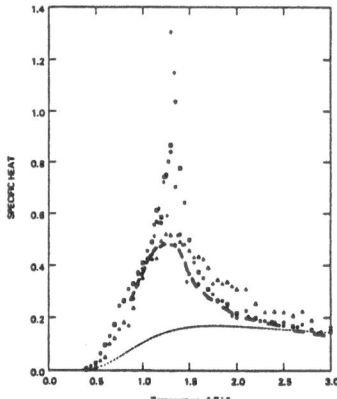

Fig. 4.4 Specific heat of the fully frustrated sc-lattice. MC   results for different
          sizes L^3 [L = 2 : (dashed), L = 4 : $\triangle$, L = 8 : $\square$, L = 20 : o   and
          L = 30 : $\lozenge$]   show the increasing maximum of   c   with increasing   L . The
          dashed line corresponds to a disordered system (ref. 136).

          first order (contrary to the   MC   data).

(b)   <u>Partially frustrated</u> (with the plaquettes of only one of the three
      orientations frustrated)

      The order parameter has two components   (n = 2 ·, XY-model)   with
      a symmetry breaking term of eighteth order irrelevant in ε-expan-
      sion. This should yield 'normal'   XY   behavior.

Diep, Lallemand and Nagai have further investiagted the fully[139] and
a different partially[140] frustrated sc-systems using   MC   techniques.
Note that in their partially frustrated model all the plaquettes of
two of the three orientations are frustrated.

From finite size analysis their estimate for the transition temperature of the fully frustrated system is:

$$T_c / |J| =   1.355 \pm 0.002   ,$$ (4.9)

close to the value of Kirkpatrick, and like him they also conclude the
transition to be second order. Their results for the specific heat

derived from dU/dT and from fluctuations of U during MC simulation at fixed temperature agree very well demonstrating thermodynamic
equilibrium (Fig. 4.5a). The corresponding internal energy U(T) is
shown in Fig. 4.5b. Below T_c there seem to be two temperature ranges

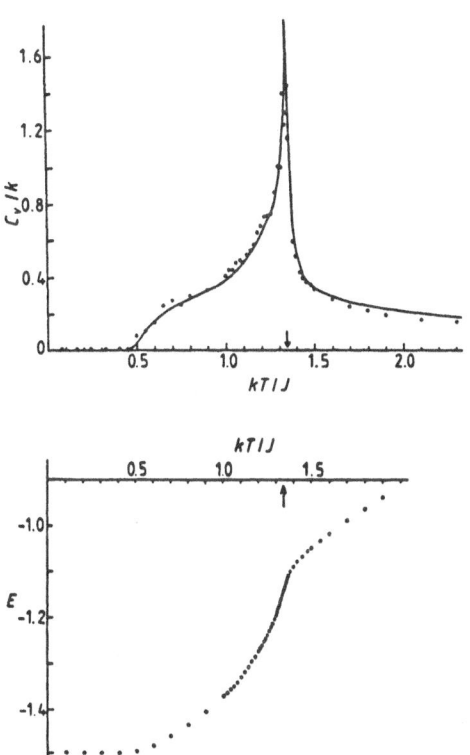

Fig. 4.5 MC results for a) (top): the specific heat c and b) (bottom): the
internal energy of the fully frustrated sc-lattice. In a) the full line
is dU/dT , the points are derived from fluctuations of U during constant temperature MC runs (ref. 139).

where the ordered phase has different properties. For $T \lesssim 0.5\ |J|$
disorder is concentrated in one sublattice containing every forth
(100) chain just as for the class of GS mentioned before. The disorder in these chains forces the chains forming the other sublattices
to remain well ordered.

For $T \gtrsim 0.5\ |J|$ the disorder is distributed more evenly on the dif-
ferent sublattices. The (continuous) crossover between both ranges

can be seen in the averaged pair correlation function for nnn-sites
along (100) directions in Fig. 4.6.

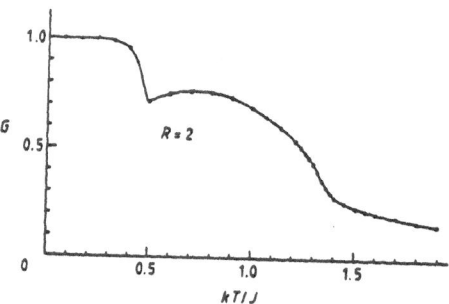

Fig. 4.6 Temperature dependence of the averaged correlation function between
nnn-sites along (100) directions (ref. 139).

Diep et al. also determined the critical exponents and find strong
deviations from the usual 3d Ising exponents[139].

For their partially frustrated model Diep et al.[140] first discuss the
GS properties. The GS exhibit 1d disorder, because any stacking
of the ferromagnetic ordered planes parallel to the nonfrustrated
plaquettes yields a GS ; similar to the fcc-case discussed in the
previous section. With a low temperature series expansions just like
that Mackenzie and Young[127] have performed they show that for low
finite temperatures the thermodynamically stable phase is the one with
alternating two planes spin up and two spin down, which they call L
phase. The other GS lead to metastable phases, the simplest being
the ferromagnetic F phase. Figure 3.7 shows the MC results for the
internal energy starting the MC runs with L respective F confi-
gurations. For both cases at temperatures up to $T \simeq 1.5 |J|$ the
agreement with the corresponding series results is very good.

With increasing temperature the metastable F phase changes into the
L phase during the simulation time used; for the system size $N = 12^3$
this happens at $T \simeq 1.7 |J|$. The transition to the paramagnetic phase
occurs at $T_c \simeq 2.5 |J|$ for the 12^3 system. A more precise value for
the infinite system as well as the order of the transition are not
known yet, although the monotonous behavior of the U(T) data sug-
gests a second order transition.

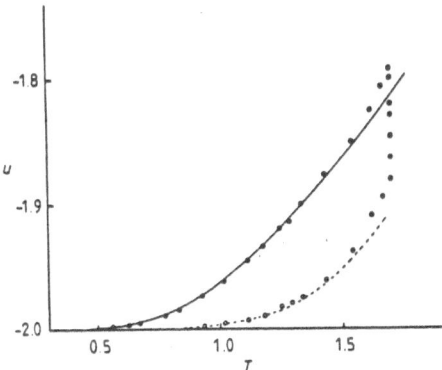

<u>Fig. 4.7</u> MC results (dots and circles) for the internal energy U starting from the stable L phase and the metastable F phase. The full and broken curves denote the series expansion results. The free energy of the L phase is lower than that of the F phase, although U is higher.

Diep and Nagai[141] have studied the effect of bond disorder for the sc Ising system, and Diep, Ghazali and Lallemand[142] the extension of the fully frustrated case to XY and Heisenberg spins, but both topics lead beyond the scope of this review.

4.3 AF Pyrochlore Model

In CsNiFeF$_6$ and CsMnFeF$_6$, both of the (modified) pyrochlore structure[143], the magnetic ions form a lattice of cornersharing tetrahedra (Fig. 4.8). This lattice is equivalent to the lattice of the B sites in spinells[144]; we shall call it pyrochlore (PC) lattice here. Steiner et al.[145] for both substances have measured the structure factor in the 110 plane by inelastic neutron scattering and, apart from weak Bragg peaks, found pronounced diffuse scattering caused by only short-ranged magnetic order (Fig. 4.9). The AF nn-interactions J_1 are strong, and the system is frustrated. MC calculations[146] of the corresponding AF Ising model show a completely monotonous internal energy for $0 < T < \infty$, which agrees very well with the energy per site obtained from a one-tetrahedron approximation[146]

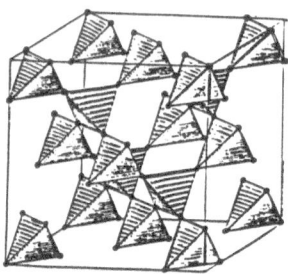

Fig. 4.8 The lattice of the magnetic ions in the pyrochlore lattice is formed by cornersharing tetrahedra.

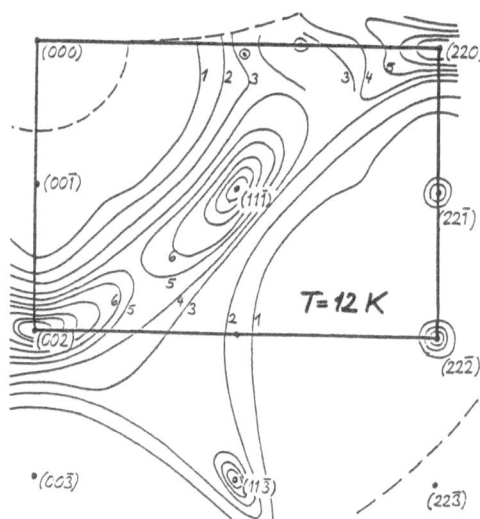

Fig. 4.9 Lines of constant structure factor in the 110 plane of $CsNiFeF_6$ (ref. 145).

$$\tilde{U}(T) \quad = \quad - 3 \ (1-x^4)/(3 + 4x + x^4) \qquad , \qquad (4.10)$$

with $x = \exp(-2|J_1|/T)$. The corresponding entropy

$$\tilde{S}(T) \quad = \quad \ln 2 + \frac{1}{2} \ln \ ((3 + 4x + x^4)/8) \qquad\qquad (4.11)$$

for $T = 0$ reproduces the Pauling approximation[39,144] for the GS entropy

$$S_0^P \quad = \quad \ln 2 + \frac{1}{2} \ln \frac{3}{8} \quad \simeq \quad 0.2925 \cdot \ln 2 \quad , \qquad (4.12)$$

in good agreement with the MC value[146]

$$S_0^{MC} \quad = \quad 0.293 \cdot \ln 2 \qquad . \qquad\qquad (4.13)$$

As the GS degeneracy of the AF PC model can be mapped exactly to that of 'cubic' ice[144,147], one can also compare with the high order series expansion of Nagle[119]:

$$S_0^N \quad = \quad (0.29577 \pm 0.00007) \cdot \ln 2 \qquad . \qquad (4.14)$$

Thus the AF PC model is much more degenerate as the frustrated fcc- and sc-lattices discussed before, and the internal energy gives no in- dication of a phase trantision.

Another indication for the absence of a phase transition is the mono- tonous decay of original LRO proportional to $\exp(-\Delta E/T)$ (see Fig. 4.10) during MC simulations for $T \geq 0.28 \ |J_1|$ [146].

As in the fcc-lattice, which also can be built from tetrahedra, the GS of the PC lattice changes at two critical fields $H_{c_1} = 2 \ |J_1|$ and $H_{c_2} = 6 \ |J_1|$. MC results for M(H) and $\chi(T) = dM/dH$ are shown in Fig. 4.11 [146]. The lower full curve in Fig. 4.11b represents the single-tetrahedron approximation

$$\tilde{\chi}(T) \quad = \quad \frac{1}{T} \cdot 2 \ (x+x^4)/(3 + 4x + x^4) \qquad , \qquad (4.15)$$

reproducing well the MC data for low temperature, a further indica- tion for the probable absence of a phase transition and for the system to remain paramagnetic down to $T = 0$ for $H = 0$.

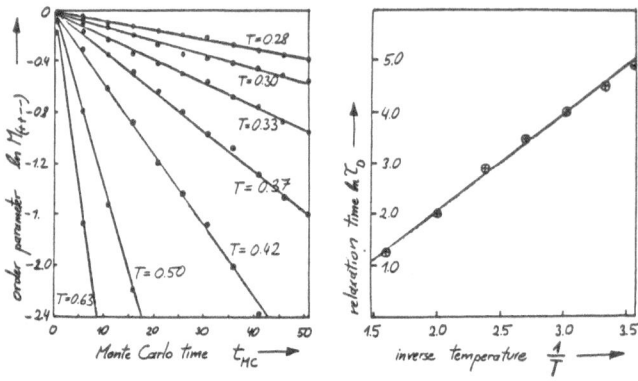

Fig. 4.10 Exponential decay of original longe range order during MC runs at different temperatures (ref. 146).

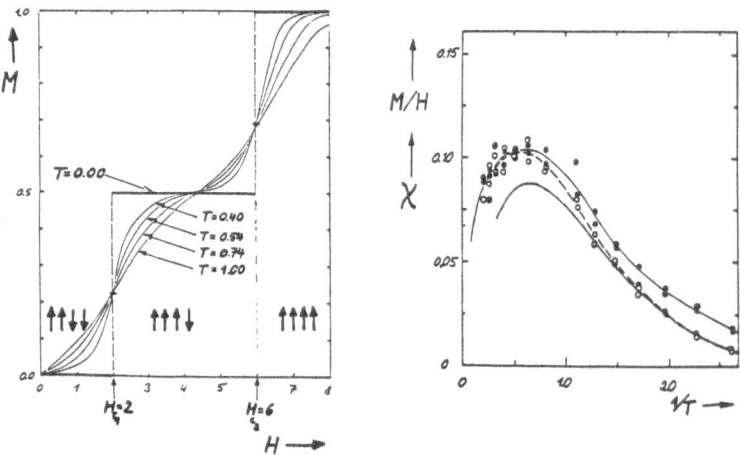

Fig. 4.11 MC results for the magnetization M(H) (a, left) , and for the susceptibility χ(T) (b, right) in the nn AF PC model (ref. 146).

Figure 4.12 shows the MC structure factor of the AF PC model in the same 110 plane as for CsNiFeF$_6$ in Fig. 4.9. Like there a very broad ridge of S(q) extends along the Brillouin zone (BZ) boundary. However, the position of the maximum does not agree with Fig. 4.9.

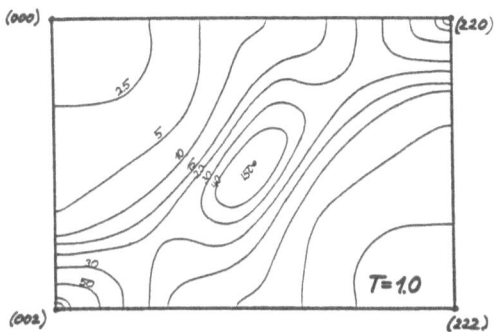

Fig. 4.12 Lines of constant MC structure factor in the nn AF PC model (ref. 146).

Introducing a weak additional nnn-interaction J$_2$ leads to better agreement with experiment (see Fig. 4.13, where J$_2$/$|$J$_1$$|$ = 0.1) .

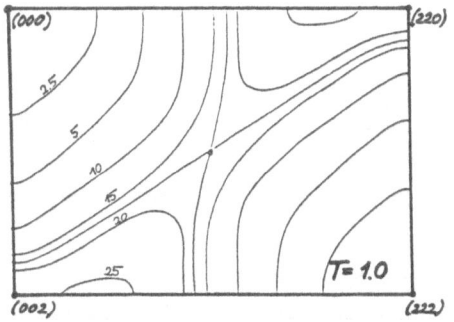

Fig. 4.13 MC structure factor as in Fig. 4.12, but with additional nnn-interactions J$_2$ = 0.1 $|$J$_1$$|$ (ref. 146).

For a more complete description of the experimental data one has to take account of the two different kinds of magnetic ions (e.g. Ni^{2+} and Fe^{3+}) distributed more or less randomly on the PC lattice. Here we do not discuss this further, as our interest is focussed on the nn AF PC model, which as far as the MC data performed can tell, remains paramagnetic down to T = 0 and which possesses a finite GS entropy per site, in contrast to the frustrated fcc- and sc-lattices.

We add that the AF PC model thus is a counterexample to the conjecture of Chui et al.[138] predicting phase transitions at finite temperature for all 3d frustrated systems.

A related Ising system with finite GS entropy when restricted to AF nn-interactions has been studied by Chui[148]. It consists of corner-sharing octahedra, forming a cubic lattice. This lattice corresponds to that of the Cu sites in $AuCu_3$ and can be decomposed into three sc-sublattices. It is easy to see that spins on one sublattice are completely free at T = 0 , when the other two are ordered ferromagnetically with opposite orientation $(S_0 > 1/3 \ln 2)$. Chui assumes the system to exhibit an ordered low temperature phase, but this remains to be proven.

4.4 ANNNI-model

The 3d ANNNI-model is shown in Fig. 4.14. The nn-interactions J_0 (in x and y direction) and J_1 (in z direction) are ferromagnetic; in z direction also nnn-spins interact antiferromagnetically via J_2 . The properties of this 3d system are quite different from the 2d case discussed in Section 3.4.1. This system has been investigated very extensively[149-151], and here we can give only a brief survey of the results. The basic phase diagram is shown in Fig. 4.15. It is derived from MC calculations[152,153], high-[154] and low-[155] temperature series expansions. For T = 0 a change of the GS from ferromagnetic to <2> occurs at κ $(= - J_2/J_1) = 1/2$, just as for d = 1 and 2 . The point (T = 0, κ = 1/2) is a frustration point, or more precisely a multiphase point with 1d degeneracy.

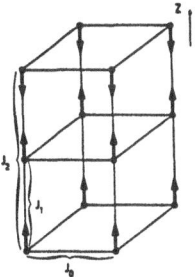

Fig. 4.14 Unit cell of the 3d ANNNI-model.

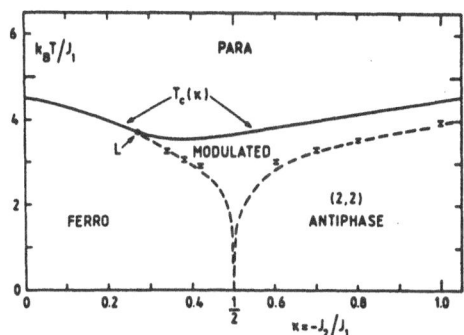

Fig. 4.15 Phase diagram of the 3d ANNNI-model (ref. 151).

Fisher and Selke[156] have found that near the multiphase point an infinite sequence of commensurate phases occurs between the F and the
<2> phases (Fig. 4.16) which all extend down to T = 0 . This is in
contrast to the 2d case where the P phase extends down to T = 0
and only one modulated M phase with continuous wavevector exists
(see. Figs. 3.38 - 3.40). The discontinuous variation of the average
wavevector $\bar{q}$ at low temperatures is shown in Fig. 4.17.

The steps in Fig. 4.17 do not form a complete "devil's staircase" in
its original meaning[157] because not all rational numbers $q/q_{<2>}$

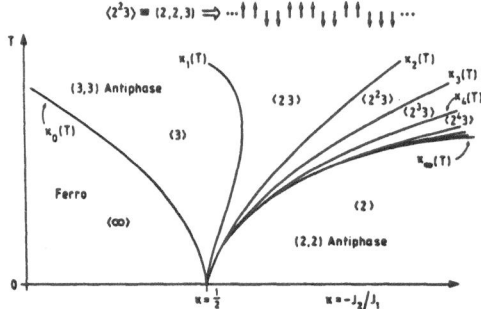

Fig. 4.16 Schematic phase diagram of the 3d ANNNI-model near the multiphase
point (ref. 156).

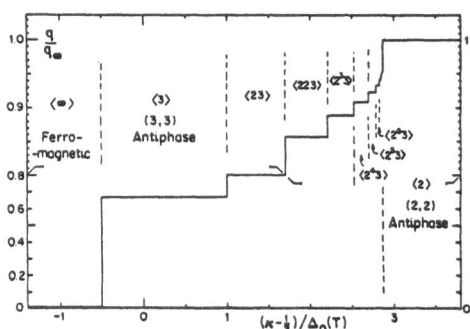

Fig. 4.17 Wavevector at low temperature as function of κ (ref. 156).

occur. Fisher and Selke have called the behavior of q near $q_{<2>}$
a "devil's top step" [158].

On the transition line to the P phase at $\kappa_L \simeq 0.27$ there is a Lifshitz point L where the F, P and M (modulated) phases meet [75,152,
154]. For $\kappa_L < \kappa < \infty$ q is expected to rise continuously from 0 to
$\pi/4$ along the M-P transition line. Selke and Duxbury [159] have investigated the complexe behavior at intermediate temperatures between
the discrete phases at low temperatures and the continuous q varia-

tion near T_c using an effectively 1d mean field (MF) approximation. They find a very complicated branch pattern, sketched roughly in
Fig. 4.18.

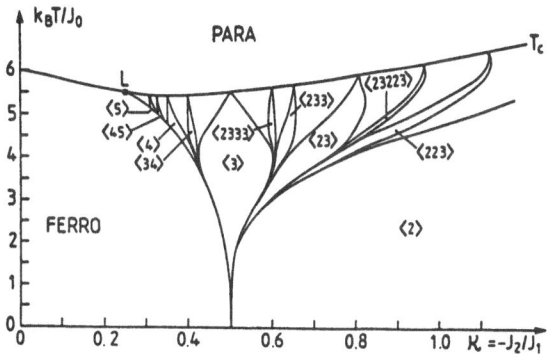

<u>Fig. 4.18</u> Mean field phase diagram of Selke and Duxbury (ref. 159).

Of special interest is the branch pattern they found for $\kappa > 1/2$
which is shown in Fig. 4.19.

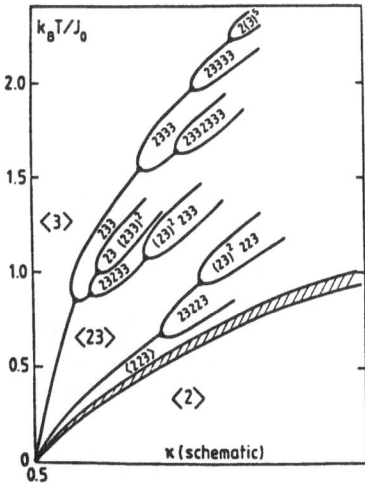

<u>Fig. 4.19</u> Branch pattern generated from combining adjacent structures (ref. 159).

The MF phase diagram (Fig. 4.18) has great similarity with the one determined by Aubry[157] for the 1d lattice model of Frank and Van der Merwe (FVdM) which describes commensurate-incommensurate transitions[160]. This connection has been discussed e.g. by Bak[149], and Axel and Aubry[161] then have shown in detail the conditions for mapping both models into each other. Further mean field results have been obtained for instance by De Simone and Stratt[162]; however, as we noted already, the ANNNI literature is too extensive to give a complete survey here.

In conclusion we just state, that the 3d ANNNI-model with competing interactions which at first glance looks so simple, exhibits a surprisingly rich phase diagram many details of which still have to be clarified.

At the end of this section we want to mention two other models with similarities to the ANNNI-model, which are formed by stacking 2d triangular or hexagon lattices with ferromagnetic nn-interactions connecting adjacent layers.

The first model has nn- and nnn-interactions within the triangular layers. The GS phase diagram is identical with the 2d case (Fig. 3.43). Nakanishi and Shiba[163] in MF approximation have found a phase diagram (Fig. 4.20) similar to the ANNNI-model, exhibiting a complete "devil's staircase" behavior.

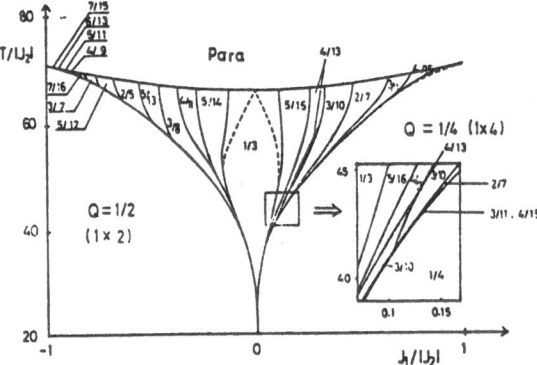

Fig. 4.20 MF phase diagram of the hexagonal model of Nakanishi et al. (ref. 163).

The model without nnn , but with AF nn-interactions within the triangular layers has been investigated for instance by Blankschtein et
al.[164] and Berker et al.[165a] within Ginzburg-Landau-Wilson theory and
RG expansions. Apart from the P phase they find two partially ordered
phases with different symmetries. However, Coppersmith[165b] has shown
that because of frustration in this model conventional low temperature
expansion as well as mean field theory using the method of ref. 163
and Landau theory are not reliable to describe the low temperature
phase.

The second model has nn- and nnn-interactions within the hexagon
layers and has been studied by Rujan[166]. He presents a GS phase
diagram (with additional four-spin interactions) where, however, the
phase boundary between the A and C phases (Fig. 3 of ref. 166) is
not correct.

4.5 fcc Four-Spin (Quartet) Model

As the last 3d Ising system we now discuss the system with four-
spin interactions J_4 on the fcc-lattice, where any four spins S_i
situated at the corners of an elementary tetrahedron of the lattice
contribute a term $J_4 \prod_{i=1}^{4} S_i$ to the Hamiltonian.

In this system at T = O there is no competition between neighboring
J_4 interactions. Nevertheless we have included this quartet model
here because like many frustrated systems it exhibits a high (1d)
GS degeneracy

$$N_g \propto N^{1/3} \quad , \tag{4.16}$$

similar to the AF fcc-model with nn-pair interactions. The quartet
model is of special interest as it is intermediate between the 2d
Ising system with global (+/-) symmetry and the 4d z_2 gauge model
with local (+/-) symmetry, both of which are self-dual[32,167].

Based on high and low temperature series expansions Wood[168] had first
assumed the system to be self-dual. As a consequence the system should
have either (I) a single phase transition at the Onsager value
$T_c^0 = 2/\ln (1+\sqrt{2}) \simeq 2.27$ or (II) a pair of transitions linked by
duality. From extended low temperature expansions Griffiths and

Wood[169] concluded case (II) to be true. MC calculations of Mouritsen et al.[170] definitely showed the existence of only one transition, but their value for T_c was far above T_c^o , excluding self-duality.

To resolve these contradictions Liebmann[171] and Pearce and Baxter[172] first proved that the quartet model in fact is self-dual, contrary to the MC result of ref. 170. Taking care of very strong hysteresis effects, in further MC calculations Liebmann[173] and then also Alcaraz et al.[174] and Mouritsen et al.[175] obtained full agreement of the MC value for T_c with T_c^o required by self-duality.

The phase transition is of first order with pronounced metastability of the low and high temperature phases beyond T_c , which was the main reason for the deviation of the original value from T_c^o .

Figure 4.21 shows the temperature dependence of the internal energy U and the entropy S , the discontinuity at T_c is unusualy strong.

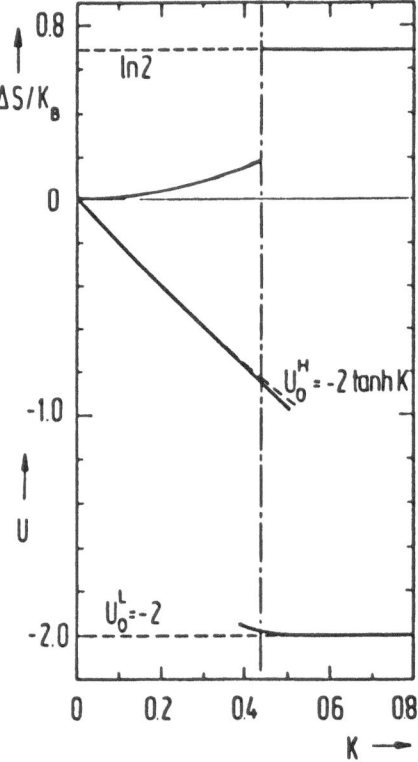

Fig. 4.21 MC results for internal energy U and entropy S of the quartet model as function of $K = J_4/T$ (ref. 173).

In the MC calculations also finite size effects are observed. Alcaraz et al.[174] found good agreement of $\beta = 1/T_c$ with the scaling law

$$\beta_0 - \beta_N \simeq \alpha \cdot N^{-1/3} \quad , \tag{4.17}$$

where for periodic boundary conditions $\alpha_p \simeq 0.1$, and for frozen ones
$\alpha_f \simeq 0.9$. $N^{1/3}$ is the linear dimension of the finite lattice with N
sites. Figure 4.22 represents the second case.

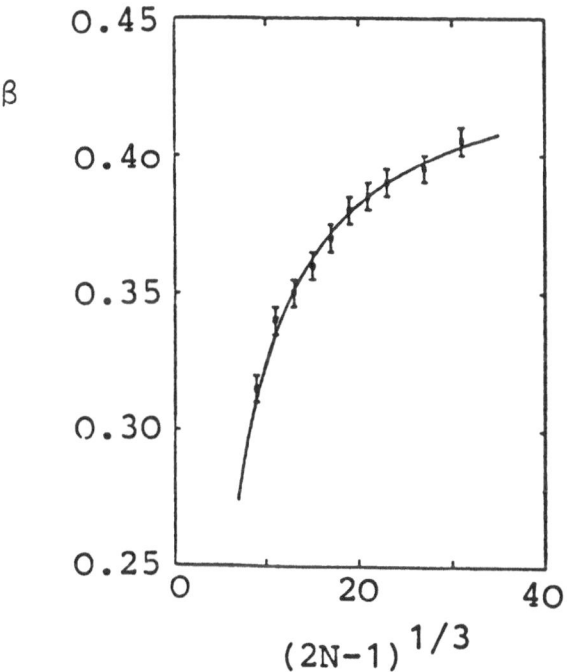

Fig. 4.22 Finite size effect on $\beta = 1/T_c$ for MC results of the quartet model
(ref. 174).

We note that in the quartet model all N_g phases are completely
equivalent, contrary to the AF fcc-model where only the LRO phases
with high symmetry are thermodynamically stable. The quartet model has
been generalized in several ways. Mouritsen et al.[175] also have studied four-spin models on the bcc and the sc-lattice and determined
the effect of additional nn-pair interactions. These lead to a tricritical point where the transition charges from first to second order.

Alcaraz et al.[176] have studied generalized fcc-quartet models for spins with Z_N symmetry (N = 2 corresponds to Ising spins). They find a first order transition for $N < N_c \simeq 5$, whereas for $N > N_c$ two continuous transitions occur which (may be) are of infinite order. Also for Z_n symmetry the system approximately remains self-dual.

Finally we note that the Ising quartet model on the other 3d close packed lattice, the hcp-lattice, also if self-dual (Liebmann[173], Alcaraz et al.[176]). The differences to the fcc- case, for instance in the free energy are minimal.

5. <u>Conclusion</u>

This book has intended to give a review of the present state of the theory of periodic frustrated Ising systems in one to three dimensions. We first discussed the relatively simple 1d systems which cannot exhibit finite temperature transitions, but show already common effects of frustration like finite groundstate (GS) entropy per site.

In two dimensions for the large group of systems with noncrossing interactions a considerable bulk of knowledge has been accumulated, because these systems can be solved exactly by transfer matrix methods. They all can be described as some kind of layered models in the sense of ref. 54. As long as they exhibit a finite temperature phase transition, it always is of the usual Ising type with the exponents of the 2d ferromagnetic case. When, however, frustration is strong enough to suppress such a transition, there are two new universality classes of systems. Systems of the first one become critical at T = 0 , where the correlation length diverges proportional to $r^{-1/2}$ (contrary to the usual 1/4 exponent). In the second class of frustrated systems the correlation length remains finite down to T = 0 , they never become critical.

Inclusion of further reaching interactions and of a magnetic field lead to a multitude of 2d systems with in general multidimensional order parameters. This causes a corresponding wealth of critical behavior like nonuniversal critical exponents, multicritical points and, for instance, commensurate-incommensurate transitions. However, most of these systems can no longer be solved exactly and one has to rely on approximate analytical and numerical methods.

This is even more true for the 3d systems, where only the simplest ones have been studied up to now. But these few ones show already very different properties. Here in the next few years many further results can be expected.

For comparison with experimental frustrated systems often the inclusion of further interactions will be necessary, as they may change the low temperature behavior drastically even when they are quite weak. Another point of interest is the generalization from Ising systems to ones with higher spin dimension (XY , Heisenberg case). This will also be very important for understanding experiments.

Finally replacing classical by quantum spins may be of importance, since frustration effects seem to be quite different[177] in both cases.

References

1. E. Ising, Z. Phys. $\underline{31}$, 253 (1925).
   H.N.V. Temperley, in: Phase Transitions and Critical Phenomena,
   Vol. 1, Eds.: C. Domb and M.S. Green, Academic Press 1972.

2. e.g.: K. Binder, in: Fundamental Problems in Statistical Mechanics,
   Ed.: E.G.D. Cohen, North Holland 1980, and: Heidelberg Colloquium
   on Spin Glasses, Eds. J.L. van Hemmen and I. Morgenstern, Lecture
   Notes in Physics $\underline{192}$, Springer 1983.

3. G. Toulouse, Commun. Phys. $\underline{2}$, 115 (1977).
   J. Vannimenus, G. Toulouse, J. Phys. $\underline{C10}$, L537 (1977).

4. A good introduction is: S.-K. Ma, Modern Theory of Critical Phenomena, Benjamin 1976.

5. from Tab. 2.1 of ref. 4.

6. D. Mukamel, S. Krinsky, Phys. Rev. $\underline{B13}$, 5065 (1976).
   S. Krinsky, D. Mukamel, Phys. Rev. $\underline{B16}$, 2313 (1977).

7. S. Alexander, P. Pincus, J. Phys. $\underline{A13}$, 263 (1980).

8. R.J. Elliott, Phys. Rev. $\underline{124}$, 346 (1961)

9. D.R. Nelson, M.E. Fisher, Ann. Phys. (Paris) $\underline{91}$, 226 (1975).

10. W. Selke, M.E. Fisher, Z. Phys. $\underline{B40}$, 71 (1980).

11. I. Peschel, Z. Phys. $\underline{B45}$, 339 (1982).

12. R.M. Hornreich, R. Liebmann, H.G. Schuster, W. Selke, Z. Phys.
    $\underline{B35}$, 91 (1979).

13. J. Stephenson, Phys. Rev. $\underline{B1}$, 4405 (1970) and Can. J. Phys.
    $\underline{48}$, 1724 (1970).

14. R. Liebmann, in: Ordering in Strongly Fluctuating Condensed Mat-
    ter Systems, p. 85, Ed.: T. Riste, NATO-ASI $\underline{B50}$, Plenum 1980.

15. R. Liebmann, unpublished.

16. M.E. Fisher, Phys. Rev. $\underline{113}$, 969 (1959).

17. D.A. Huse, M.E. Fisher, J.M. Yeomans, Phys. Rev. $\underline{B23}$, 180 (1981).

18. Graphical representation of the results of ref. 17 for $n = 9$.

19. B.G.S. Doman, J.K. Williams, J. Phys. $\underline{C15}$, 1693 (1982).

20. J.K. Williams, J. Phys. $\underline{C14}$, 4095 (1981).

21. F. Matsubara, Prog. Theor. Phys. $\underline{51}$, 1694 (1974).

22. D.P. Landau, M. Blume, Phys. Rev. $\underline{B13}$, 287 (1976).

23. J.F. Fernandez, Phys. Rev. $\underline{B16}$, 5125 (1977).

24. B. Derrida, J. Vannimenus, Y. Pomeau, J. Phys. $\underline{C11}$, 4749 (1978).

25. M. Puma, J.F. Fernandez, Phys. Rev. $\underline{B18}$, 1391 (1978).

26. N.D. Mermin, H. Wagner, Phys. Rev. Lett. $\underline{17}$, 1133 (1966).

27. H.A. Kramers, G.H. Wannier, Phys. Rev. $\underline{60}$, 252 (1941).

28. I. Syozi, Prog. Theor. Phys. $\underline{6}$, 306 (1951).

29. L. Onsager, Phys. Rev. $\underline{65}$, 117 (1944).

30. I. Syozi, in: Phase Transitions and Critical Phenomena, Vol. I, Eds.: C. Domb and M.S. Green, Academic Press 1972.

31. M. Tokunaga, T. Matsubara, Prog. Theor. Phys. $\underline{35}$, 581 (1966).

32. F.J. Wegner, J. Math. Phys. $\underline{12}$, 2259 (1971).

33. E. Fradkin, B.A. Huberman, S.H. Shenker, Phys. Rev. $\underline{B18}$, 4789 (1978).

34. M.E. Fisher, Phys. Rev. $\underline{113}$, 969 (1959).

35. G. Forgacs, Phys. Rev. $\underline{B22}$, 4473 (1980).

36. I. Peschel, Z. Phys. $\underline{B45}$, 339 (1982).

37. G.H. Wannier, Phys. Rev. $\underline{79}$, 357 (1950), Phys. Rev. $\underline{B7}$, 5017 ($\underline{E}$) (1973).

38. R.M.F. Houtappel, Physica $\underline{16}$, 425 (1950).

39. L. Pauling: The Nature of the Chemical Bond, Cornell Univ. Press 1938.

40. M.E. Fisher, in: Critical Phenomena, Ed.: M.S. Green, Academic Press 1971.

41. M.E. Fisher, M.N. Barber, Phys. Rev. Lett. $\underline{28}$, 1516 (1972).

42. M.N. Barber, in: Phase Transitions and Critical Phenomena, Vol. $\underline{10}$, Eds.: C. Domb and J.L. Lebowitz, Academic Press

43. V.G. Vaks, M.B. Geilikman, Zh. Eksp. Teor. Fiz. $\underline{60}$, 330 (1971); or: Sov. Phys. JETP $\underline{33}$, 179 (1971).

44. J. Stephenson, J. Math. Phys. $\underline{11}$, 420 (1970).

45. J. Stephenson, J. Math. Phys. $\underline{11}$, 413 (1970).

46. C.A. Hurst, H.S. Green: Order-Disorder Phenomena, Wiley Inter-science 1964.

47. M.J. Stephen, L. Mittag, J. Math. Phys. $\underline{13}$, 1944 (1972).

48. J.M. Luttinger, J. Math. Phys. $\underline{4}$, 1154 (1963).
D.C. Mattis, E.H. Lieb, J. Math. Phys. $\underline{6}$, 304 (1965).
A. Theumann, J. Math. Phys. $\underline{8}$, 2460 (1967).

49. I. Peschel, V.J. Emery, Z. Phys. $\underline{B43}$, 241 (1981).

50. R. Glauber, J. Math. Phys. $\underline{4}$, 294 (1963).

51. J.C. Kimball, J. Stat. Phys. $\underline{22}$, 289 (1979).

52. R. Rujan, J. Stat. Phys. $\underline{29}$, 231 (1982).

53. R. Rujan, J. Stat. Phys. $\underline{29}$, 247 (1982); $\underline{34}$, 615 (1984).

54. W.F. Wolff, J. Zittartz, Z. Phys. $\underline{B47}$, 341 (1982); $\underline{B49}$, 229 (1982); $\underline{B50}$, 131 (1983).

55. V.G. Vaks, A.I. Larkin, Y.N. Ovchinnikov, J. Exptl. Theor. Phys. $\underline{49}$, 1180 (1965); or Sov. Phys. JETP $\underline{22}$, 820 (1966).

56. W. Kränzig: Ferroelectrics and Antiferroelectrics, Academic Press 1957.
J. Burfoot: Ferroelectrics, Van Nostrand 1965.

57. E.H. Fradkin, T.P. Eggarter, Phys. Rev. $\underline{A14}$, 495 (1976).

58. J. Villain, J. Phys. $\underline{C10}$, 1717 (1977).

59. G. André, R. Bidaux, J.-P. Carton, R. Conte, L. De Seze, J. Phys. (Paris) $\underline{40}$, 479 (1979).

60. M.E. Fisher, Phys. Rev. $\underline{124}$, 1664 (1961).

61. M. Gabay, J. Phys. Lett. (Paris) $\underline{41}$, 427 (1980).

62. In ref. 54 the other papers are cited.

63. W.F. Wolff, P. Hoever, J. Zittartz, Z. Phys. $\underline{B42}$, 259 (1981).

64. W.F. Wolff, J. Zittartz, Z. Phys. B44, 109 (1981).

65. A. Süto, Z. Phys. B44, 121 (1981).

66. W.F. Wolff, J. Zittartz, Z. Phys. B49, 139 (1982).

67. M.H. Waldor, W.F. Wolff, J. Zittartz, Phys. Lett. 106A, 261 (1984); Z. Phys. B59, 43 (1985).

68. K. Kano, S. Naya, Prog. Theor. Phys. Jpn. 10, 158 (1953).

69. M.B. Geilikman, Zh. Eksp. Teor. Fiz. 66, 1166 (1974); or Sov. Phys. JETP 39, 570 (1974).

70. A. Süto, Helv. Phys. Acta 54, 191 + 201 (1981).

71. P. Hoever, W.F. Wolff, J. Zittartz, Z. Phys. B41, 43 (1981), W. Wolff, thesis, Köln 1981.

72. W. Selke, M.E. Fischer, Z. Phys. B40, 71 (1980).
    W. Selke, Z. Phys. B43, 335 (1981).
    M.N. Barber, W. Selke, J. Phys. A15, L617 (1982).

73. (a) J. Kroemer, W. Pesch, J. Phys. A15, L25 (1982).
    (b) W. Pesch, J. Kroemer, Z. Phys. B59, 317 (1985).

74. J. Villain, P. Bak, J. Phys. (Paris) 42, 657 (1981).

75. R.M. Hornreich, M. Luban, S. Shtrikman, Phys. Rev. Lett. 35, 1678 (1975).
    R.M. Hornreich, J. Magn. Magn. Mater. 15-18, 387 (1980).

76. E. Müller-Hartmann, J. Zittartz, Z. Phys. B27, 261 (1977).

77. J.M. Kosterlitz, D.J. Thouless, J. Phys. C6, 1181 (1973).

78. R. Bidaux, L. de Seze, J. Phys. (Paris) 42, 371 (1981).

79. I. Morgenstern, Phys. Rev. B26, 5296 (1982).

80. M. Doukouré, D. Gignoux, J. Magn. Magn. Mater. 30, 111 (1982).

81. e.g.: A.N. Berker, S. Ostlund, F.A. Putnam, Phys. Rev. B17, 3650 (1978),
    R.J. Birgeneau, E.M. Hammons, P. Heiny, P.W. Stephens, P.M. Horn, in: Ordering in Two Dimensions, Ed. S.K. Sinha, Elsevier 1980.

82. B.D. Metcalf, Phys. Lett. 45A, 1 (1973).

83. Y. Tanaka, N. Uryu, Prog. Theor. Phys. 55, 1356 (1976).

84. M. Kaburagi, J. Kanamori, J. Phys. Soc. Jpn. $\underline{44}$, 718 (1978).

85. J. Oitmaa, J. Phys. $\underline{A15}$, 573 (1982).

86. E. Domany, M. Schick, J.S. Walker, R.B. Griffiths, Phys. Rev. $\underline{B18}$, 2209 (1978), also ref. 8.

87. J.V. José, L.P. Kadanoff, S. Kirkpatrick, D.R. Nelson, Phys. Rev. $\underline{B16}$, 1217 (1977).

88. D.P. Landau, Phys. Rev. $\underline{B27}$, 5604 (1983).

89. K. Wada, T. Ishikawa, J. Phys. Soc. Jpn. $\underline{52}$, 1774 (1983).

90. M. Kaburagi, T. Tonegawa, J. Kanamori, J. Magn. Magn. Mater. $\underline{31}$, 1037 (1983).

91. P.A. Slotte, P.C. Hemmer, J. Phys. $\underline{C17}$, 4645 (1984).

92. M. Kaburagi, J. Kanamori, Jpn. J. Appl. Phys. Suppl. $\underline{2}$, 145 (1974).

93. S. Alexander, Phys. Lett. $\underline{A54}$, 353 (1975).

94. M. Schick, J.S. Walker, M. Wortis, Phys. Lett. $\underline{A58}$, 479 (1976).

95. J. Doczi-Reger, P.C. Hemmer, Physica $\underline{108A}$, 531 (1981).

96. K. Nakanishi, H. Shiba, J. Phys. Soc. Jpn. $\underline{51}$, 2089 (1982).

97. B. Mihura, D.P. Landau, Phys. Rev. Lett. $\underline{38}$, 977 (1977).

98. J. Kanamori, J. Phys. Soc. Jpn. $\underline{53}$, 250 (1984).

99. M.E. Fisher, W. Selke, Phil. Trans. Roy. Soc. London $\underline{A302}$, 1 (1981).

100. J.G. Dash: Films on Solid Surfaces, Academic Press 1975.

101. N.W. Dalton, D.W. Wood, J. Math. Phys. $\underline{10}$, 1271 (1969).

102. S. Miyashita, J. Phys. Soc. Jpn. $\underline{52}$, 780 (1983).

103. S. Fujiki, K. Shutoh, S. Katsura, J. Phys. Soc. Jpn. $\underline{53}$, 1371 (1984).

104. J.M.J. van Leeuwen, Phys. Rev. Lett. $\underline{34}$, 1056 (1975).

105. M. Nauenberg, B. Nienhuis, Phys. Rev. Lett. $\underline{33}$, 944 (1974).

106. M.N. Nightingale, Phys. Lett. A59, 486 (1977).

107. L.P. Kadanoff, F.J. Wegner, Phys. Rev. B4, 3989 (1971).

108. R.J. Baxter, Phys. Rev. Lett. 26, 832 (1971).

109. M.N. Barber, J. Phys. A12, 679 (1979).

110. (a) J. Kanamori, Prog. Theor. Phys. 35, 16 (1966).
     (b) U. Brandt, Z. Phys. B53, 283 (1983).

111. P.D. Beale, J. Phys. A17, L335 (1984).

112. See citations in ref. 111.

113. K. Binder, D.P. Landau, Phys. Rev. B21, 1941 (1980).

114. D.C. Rapaport, C. Domb, J. Phys. C4, 2684 (1971).
     N.W. Dalton, D.W. Wood, J. Math. Phys. 10, 1271 (1969).

115. K.R. Subbaswamy, G.D. Mahan, Phys. Rev. Lett. 37, 642 (1976).

116. R.J. Baxter, I.G. Enting, S.K. Tsang, J. Stat. Phys. 22, 465
     (1980).

117. J. Doczi-Reger, P.C. Hemmer, phys. stat. sol. (b) 116, K67
     (1983).

118. E.H. Lieb, F.Y. Wu, in: Phase Transitions and Critical Phenomena,
     Eds.: C. Domb, M.S. Green, Academic Press 1972.

119. J.F. Nagle, J. Math. Phys. 7, 1484 + 1492 (1966).

120. E.H. Lieb, Phys. Rev. 162, 162 (1967); Phys. Rev. Lett. 18, 1046
     (1967).

121. L. Pauling, J. Am. Chem. Soc. 57, 2680 (1935) and ref. 39.

122. R.J. Baxter, F.Y. Wu, Phys. Rev. Lett. 31, 1294 (1973).

123. J. Doczi-Reger, P.C. Hemmer, Physica 109A, 541 (1981).

124. A. Danielian, Phys. Rev. 133, A1344 (1964).

125. D.D. Betts, C.J. Elliott, Phys. Lett. 18, 18 (1965).

126. J. Slawny, J. Stat. Phys. 20, 711 (1979).

127. N.D. Mackenzie, A.P. Young, J. Phys. C14, 3927 (1981).

128. R. Bidaux, J.P. Carton, R. Conte, J. Villain, in: Disordered Systems and Localization, Eds.: C. Catellani et al., Lecture Notes in Physics; Vol. 149, Springer 1981.

129. M.H. Phani, J.L. Lebowitz, M.H. Kalos, C.C. Tsai, Phys. Rev. Lett. 42, 577 (1979).

130. K. Binder, Phys. Rev. Lett. 45, 811 (1980).

131. K. Binder, J.L. Lebowitz, M.H. Kalos, M.H. Phani, Acta Met. 29, 1655 (1981).

132. K. Binder, Z. Phys. B45, 61 (1981).

133. E. Domany, Y. Shnidman, D. Mukamel, cited as preprint in ref. 132

134. B. Derrida, Y. Pomeau, G. Toulouse, J. Vannimenus, J. Phys. (Paris) 40, 617 (1979); 41, 213 (1980).

135. D. Blankschtein, M. Ma, A.N. Berker, Phys. Rev. B30, 1362 (1984).

136. S. Kirkpatrick, in the same proceedings as ref. 128.

137. G. Bhanot, M. Creutz, Phys. Rev. B22, 3370 (1980); see also A. Weinkauf, J. Zittartz, Z. Phys. B45, 223 (1982).

138. S.T. Chui, G. Forgacs, D.M. Hatch, Phys. Rev. B25, 6952 (1982).

139. H.T. Diep, P. Lallemand, O. Nagai, J. Phys. C18, 1067 (1985).

140. H.T. Diep, O. Nagai, J. Phys. C17, 1357 (1984).

141. H.T. Diep, O. Nagai, J. Phys. C18, 369 (1985).

142. H.T. Diep, A. Ghazali, P. Lallemand, appears in: J. Phys. C.

143. D. Babel, Z. anorg. allg. Chem. 387, 161 (1972).

144. P.W. Anderson, Phys. Rev. 102, 1008 (1956).

145. M. Steiner, S. Krasnicki, H. Dachs, R.v. Wallpach, J. Phys. Soc. Jpn. Suppl. 52, 173 (1983).

146. R. Liebmann, unpublished.

147. J. Villain, Z. Phys. B33, 31 (1979).

148. S.T. Chui, Phys. Rev. B15, 307 (1977).

149. P. Bak, Rep. Prog. Phys. 45, 587 (1982).

150. M.E. Fisher, D.A. Huse, in: Melting, Localization and Chaos, Eds.: R.K. Kalia, P. Vashishta, Elsevier 1982.

151. W. Selke, in: Modulated Structure Materials, Ed.: T. Tsakalakos, 1983.

152. W. Selke, Z. Phys. B29, 133 (1978).

153. W. Selke, M.E. Fisher, Phys. Rev. B20, 257 (1979); J. Magn. Magn. Mater. 15-18, 403 (1980).

154. S. Redner, H.E. Stanley, J. Phys. C10, 4765 (1977); Phys. Rev. B16, 4901 (1977).

155. M.E. Fisher, W. Selke, Phil. Trans. Roy. Soc. A302, 1 (1981).

156. M.E. Fisher, W. Selke, Phys. Rev. Lett. 44, 1502 (1980).

157. S. Aubry: Soliton and Condensed Matter Physics, Springer 1978, p. 264.
E. Allroth, H. Müller-Krumbhaar, Phys. Rev. A27, 1575 (1983).

158. M.E. Fisher, J. Appl. Phys. 52, 2014 (1981).

159. W. Selke, P.M. Duxbury, Z. Phys. B57, 49 (1984).
P.M. Duxbury, W. Selke, J. Phys. A16, 1741 (1983).

160. F.C. Frank, J.H. Van der Merwe, Proc. Roy. Soc. London 198, 205 + 216 (1949).

161. F. Axel, S. Aubry, J. Phys. C14, 5433 (1981).

162. T. De Simone, R.M. Stratt, Phys. Rev. B32, 1537 + 1549 (1985).

163. K. Nakanishi, H. Shiba, J. Phys. Soc. Jpn. 51, 2089 (1982).
K. Nakanishi, J. Phys. Soc. Jpn. 52, 2449 (1983).

164. D. Blankschtein, M. Ma. A.N. Berker, G.S. Grest, C.M. Soukoulis, Phys. Rev. B29, 5250 (1984).

165. (a) N. Berker, G.S. Grest, C.M.Soukoulis, D. Blankschtein, M. Ma, J. Appl. Phys. 55, 2416 (1984).
(b) S.N. Coppersmith, Phys. Rev. B32, 1584 (1985).

166. P. Rujan, J. Stat. Phys. 29, 231 (1982).

167. F.J. Wegner, Physica 68, 570 (1973).

168. D.W. Wood, J. Phys. C5, L181 (1972).

169. H.P. Griffiths, D.W. Wood, J. Phys. C6, 2533 (1973).

170. O.G. Mouritsen, S.J. Knak Jensen, B. Frank, Phys. Rev. $\underline{B23}$, 976 (1981); $\underline{B24}$, 347 (1981).

171. R. Liebmann, Phys. Lett. $\underline{85A}$, 59 (1981).

172. P.A. Pearce, R.J. Baxter, Phys. Rev. $\underline{B24}$, 5295 (1981).

173. R. Liebmann, Z. Phys. $\underline{B45}$, 243 (1982).

174. F.C. Alcaraz, L. Jacobs, R. Savit, Phys. Lett. $\underline{89A}$, 49 (1982).

175. O.G. Mouritsen, B. Frank, D. Mukamel, Phys. Rev. $\underline{B27}$, 3018 (1983).

176. F.C. Alcaraz, L. Jacobs, R. Savit, J. Phys. $\underline{A16}$, 175 (1982).

177. L.G. Marland, D.D. Betts, Phys. Rev. Lett. $\underline{43}$, 1618 (1979).

Acknowledgements

The author thanks Prof. Dr. H.G. Schuster and the other members of the Institute for Theoretical Physics, University Frankfurt, FRG, for stimulating discussions.

He also thanks Prof. Dr. K. Binder, then at the IFF, KFA Jülich, FRG, for introducing him to Monte Carlo calculations, and Prof. Dr. M. Steiner, HMI Berlin, for collaboration concerning the investigation of pyrochlore systems.

Part of this work has been supported by the Deutsche Forschungsgemeinschaft through the Sonderforschungsbereich 65 Frankfurt-Darmstadt. The manuscript was finished at the Max-Planck-Institut für Festkörperforschung, Stuttgart, FRG. For the very efficient typing I thank Mrs. Hanna-Daoud, Institute for Theoretical Physics, Technische Hochschule Darmstadt, FRG.

O. Bratteli, D. W. Robinson

Operator Algebras and Quantum Statistical Mechanics 1

C* and W*-Algebras. Symmetry Groups. Decomposition of States

1979. 1 figure. XII, 500 pages. (Texts and Monographs in Physics). ISBN 3-540-09187-4

Contents: Introduction. – C*-Algebras and von Neumann Algebras. – Groups, Semigroups, and Generators. – Decomposition Theory. – References. – List of Symbols. – Subject Index.

O. Bratteli, D. W. Robinson

Operator Algebras and Quantum Statistical Mechanics 2

Equilibrium States. Models in Quantum Statistical Mechanics

1981. XI, 505 pages. (Texts and Monographs in Physics) ISBN 3-540-10381-3

Contents: States in Quantum Statistical Mechanics: Introduction. Continuous Quantum Systems. I. KMS States. Stability and Equilibrium. Notes and Remarks. – Models of Quantum Statistical Mechanics: Introduction. Quantum Spin Systems. Continuous Quantum Systems. II. Conclusion. – Notes and Remarks. – References. – Books and Monographs. – Articles. – List of Symbols. – Subject Index. – Corrigenda to Volume 1.

O. G. Mouritsen

Computer Studies of Phase Transitions and Critical Phenomena

1984. 79 figures. XII, 200 pages. (Springer Series in Computational Physics). ISBN 3-540-13397-6

Contents: Introduction. – Computer Methods in the Study of Phase Transitions and Critical Phenomena. – Monte Carlo Pure-model Calculations. – Testing Modern Theories of Critical Phenomena. – Numerical Experiments. Bibliography. – Subject Index.

Springer-Verlag
Berlin Heidelberg
New York Tokyo

Lecture Notes in Physics

Vol. 214: H. Moraal, Classical, Discrete Spin Models. VII, 251 pages. 1984.

Vol. 215: Computing in Accelerator Design and Operation. Proceedings, 1983. Edited by W. Busse and R. Zelazny. XII, 574 pages. 1984.

Vol. 216: Applications of Field Theory to Statistical Mechanics. Proceedings, 1984. Edited by L. Garrido. VIII, 352 pages. 1985.

Vol. 217: Charge Density Waves in Solids. Proceedings, 1984. Edited by Gy. Hutiray and J. Sólyom. XIV, 541 pages. 1985.

Vol. 218: Ninth International Conference on Numerical Methods in Fluid Dynamics. Edited by Soubbaramayer and J. P. Boujot. X, 612 pages. 1985.

Vol. 219: Fusion Reactions Below the Coulomb Barrier. Proceedings, 1984. Edited by S.G. Steadman. VII, 351 pages. 1985.

Vol. 220: W. Dittrich, M. Reuter, Effective Lagrangians in Quantum Electrodynamics. V, 244 pages. 1985.

Vol. 221: Quark Matter '84. Proceedings, 1984. Edited by K. Kajantie. VI, 305 pages. 1985.

Vol. 222: A. García, P. Kielanowski, The Beta Decay of Hyperons. Edited by A. Bohm. VIII, 173 pages. 1985.

Vol. 223: H. Saller, Vereinheitlichte Feldtheorien der Elementarteilchen. IX, 157 Seiten. 1985.

Vol. 224: Supernovae as Distance Indicators. Proceedings, 1984. Edited by N. Bartel. VI, 226 pages. 1985.

Vol. 225: B. Müller, The Physics of the Quark-Gluon Plasma. VII, 142 pages. 1985.

Vol. 226: Non-Linear Equations in Classical and Quantum Field Theory. Proceedings, 1983/84. Edited by N. Sanchez. VII, 400 pages. 1985.

Vol. 227: J.-P. Eckmann, P. Wittwer, Computer Methods and Borel Summability Applied to Feigenbaum's Equation. XIV, 297 pages. 1985.

Vol. 228: Thermodynamics and Constitutive Equations. Proceedings, 1982. Edited by G. Grioli. V, 257 pages. 1985.

Vol. 229: Fundamentals of Laser Interactions. Proceedings, 1985. Edited by F. Ehlotzky. IX, 314 pages. 1985.

Vol. 230: Macroscopic Modelling of Turbulent FLows. Proceedings, 1984. Edited by U. Frisch, J. B. Keller, G. Papanicolaou and O. Pironneau. X, 360 pages. 1985.

Vol. 231: Hadrons and Heavy Ions. Proceedings, 1984. Edited by W. D. Heiss. VII, 458 pages. 1985.

Vol. 232: New Aspects of Galaxy Photometry. Proceedings, 1984. Edited by J.-L. Nieto. XIII, 350 pages. 1985.

Vol. 233: High Resolution in Solar Physics. Proceedings, 1984. Edited by R. Muller. VII, 320 pages. 1985.

Vol. 234: Electron and Photon Interactions at Intermediate Energies. Proceedings, 1984. Edited by D. Menze, W. Pfeil and W. J. Schwille. VII, 481 pages. 1985.

Vol. 235: G. E. A. Meier, F. Obermeier (Eds.), Flow of Real Fluids. VIII, 348 pages. 1985.

Vol. 236: Advanced Methods in the Evaluation of Nuclear Scattering Data. Proceedings, 1985. Edited by H. J. Krappe and R. Lipperheide. VI, 364 pages. 1985.

Vol. 237: Nearby Molecular Clouds. Proceedings, 1984. Edited by G. Serra. IX, 242 pages. 1985.

Vol. 238: The Free-Lagrange Method. Proceedings, 1985. Edited by M. J. Fritts, W. P. Crowley and H. Trease. IX, 313 pages. 1985.

Vol. 239: Geometrics Aspects of the Einstein Equations and Integrable Systems. Proceedings, 1984. Edited by R. Martini. V, 344 pages. 1985.

Vol. 240: Monte-Carlo Methods and Applications in Neutronics, Photonics and Statistical Physics. Proceedings, 1985. Edited by R. Alcouffe, R. Dautray, A. Forster, G. Ledanois and B. Mercier. VIII, 483 pages. 1985.

Vol. 241: Numerical Simulation of Combustion Phenomena. Proceedings, 1985. Edited by R. Glowinski, B. Larrouturou and R. Temam. IX, 404 pages. 1985.

Vol. 242: Exactly Solvable Problems in Condensed Matter and Relativistic Field Theory. Proceedings, 1985. Edited by B. S. Shastry, S. S. Jha and V. Singh. V, 318 pages. 1985.

Vol. 243: Medium Energy Nucleon and Antinucleon Scattering. Proceedings, 1985. Edited by H. V. von Geramb. IX, 576 pages. 1985.

Vol. 244: W. Dittrich, M. Reuter, Selected Topics in Gauge Theories. V, 315 pages. 1986.

Vol. 245: R. Kh. Zeytounian, Les Modèles Asymptotiques de la Mécanique des Fluides I. IX, 260 pages. 1986.

Vol. 246: Field Theory, Quantum Gravity and Strings. Proceedings, 1984/85. Edited by H. J. de Vega and N. Sánchez. VI, 381 pages. 1986.

Vol. 247: Nonlinear Dynamics Aspects of Particle Accelerators. Proceedings, 1985. Edited by J.M. Jowett, M. Month and S. Turner. VIII, 583 pages. 1986.

Vol. 248: Quarks and Leptons. Proceedings, 1985. Edited by C.A. Engelbrecht. X, 417 pages. 1986.

Vol. 249: Trends in Applications of Pure Mathematics to Mechanics. Proceedings, 1985. Edited by E. Kröner and K. Kirchgässner. VIII, 523 pages. 1986.

Vol. 250: Lie Methods in Optics. Proceedings 1985. Edited by J. Sánchez Mondragón and K. B. Wolf. XIV, 249 pages. 1986.

Vol. 251: R. Liebmann, Statistical Mechanics of Periodic Frustrated Ising Systems. VII, 142 pages. 1986.